THE
KNOWLEDGE
MACHINE

ALSO BY MICHAEL STREVENS

Thinking Off Your Feet: How Empirical Psychology
Vindicates Armchair Philosophy

Tychomancy: Inferring Probability from Causal Structure

Depth: An Account of Scientific Explanation

Bigger than Chaos: Understanding Complexity through Probability

THE
KNOWLEDGE
MACHINE

How Irrationality Created Modern Science

MICHAEL STREVENS

WITHDRAWN

LIVERIGHT PUBLISHING CORPORATION

A Division of W. W. Norton & Company

Independent Publishers Since 1923

NEW YORK · LONDON

Frontispiece: Eduardo Paolozzi, *Newton*, 1995.
After Blake's *Newton* (Figure 8.4).
The knowledge machine personified, or the scientific persona mechanized.

For information about permission to reproduce selections from this book,
write to Permissions, Liveright Publishing Corporation, a division of
W. W. Norton & Company, Inc., 500 Fifth Avenue, New York, NY 10110

For information about special discounts for bulk purchases, please contact
W. W. Norton Special Sales at specialsales@wwnorton.com or 800-233-4830

Manufacturing by LSC Communications, Harrisonburg
Book design by Ellen Cipriano
Production manager: Lauren Abbate

Library of Congress Cataloging-in-Publication Data

Names: Strevens, Michael, author.
Title: The knowledge machine : how irrationality created modern science /
Michael Strevens.
Description: First edition. | New York : Liveright Publishing Corporation,
[2020] | Includes bibliographical references and index.
Identifiers: LCCN 2020018496 | ISBN 9781631491375 (hardcover) |
ISBN 9781631491382 (epub)
Subjects: LCSH: Science—Methodology. | Science—Philosophy. |
Science—History. | Irrationality (Philosophy) | Knowledge, Theory of.
Classification: LCC Q175 .S865 2020 | DDC 500—dc23
LC record available at https://lccn.loc.gov/2020018496

Liveright Publishing Corporation, 500 Fifth Avenue, New York, N.Y. 10110
www.wwnorton.com

W. W. Norton & Company Ltd., 15 Carlisle Street, London W1D 3BS

1 2 3 4 5 6 7 8 9 0

Here grows the cure of all, this fruit divine,
Fair to the eye, inviting to the taste,
Of virtue to make wise: what hinders then
To reach, and feed at once both body and mind?

MILTON, *PARADISE LOST*,
BOOK 9, 776–779

CONTENTS

IV. SCIENCE NOW

THE

KNOWLEDGE
MACHINE

The Knowledge Machine

ତ୍ୱର

Why is science so powerful?
Why did it take so long to arrive?

WERE YOU TO BE TRANSPORTED to some randomly chosen place and time in human history, you would likely find yourself gathering pinhead-sized grains, hunting dangerous game with a sharpened stick, and living in a damp, unfurnished hollow.

If you were very lucky, however, you might wake up in the sandals of a wealthy citizen of the Greek world around the time of Alexander the Great. From this position of privilege you could enjoy just about every cultural invention that makes life worth living today. You could delight in the poetry of Homer and Sappho, visit the theater to relish *Oedipus Rex* and other masterpieces of ancient drama, hire a musician to serenade your friends after dinner. You could live in cities regulated by law and a system of courts, shaped by the architects and sculptors who built some of the seven wonders of the world, and governed in accordance with political models that have lasted to this day: monarchy, oligarchy, sweet democracy. If you had the aptitude and the inclination, you could undertake advanced studies in geometry or philosophy.

Yet—you would soon notice a few things missing even from this cultural Elysium. X-rays and MRIs, travel at speeds faster than the swiftest horse, voices and video coverage of world events streaming through the thyme-scented Mediterranean air are nowhere to be found. Most egregiously absent of all is the thing that makes our own advanced medicine, transport, and telecommunication possible: the knowledge-producing machine that we call modern science.

Civilization stretches back millennia. But the machine has been around only a few hundred years. What took so long?

Among the ancients there was no lack of desire to understand the workings of the world. Around 580 BCE, the Greek philosopher Thales gazed out from the port city of Miletus into the Aegean blue, into the summer haze where sea merges imperceptibly with sky, and proposed that everything is ultimately made of water. His student Anaximenes disagreed: the fundamental stuff, he said, is air. Heraclitus, who lived in Sicily a few decades later, suggested fire. Back in Miletus, Anaximander— another student of Thales about whom we know equally little—had meanwhile hypothesized that all things are composed of invisible stuff of unlimited potential that he called *apeiron*, or "the boundless."

Though these thinkers, their contemporaries, and their successors— Chinese scholars, Islamic doctors, medieval European monks—argued ingeniously for their points of view, no one of their ideas could gain a foothold over the others. Searching for the deep structure of nature, they contributed immeasurably to humanity's stock of brilliant and original hypotheses, but to its stock of knowledge, scarcely at all.

The reason is straightforward. Although premodern inquirers into nature sometimes hit on the right idea, they had little ability to distinguish it from its rivals. By the time of the collapse of the Western Roman Empire in the fifth century CE, almost every possible hypothesis about the relation between the earth, the planets, and the sun had been proposed: that the planets and the sun revolved around a fixed earth, that the earth and the planets revolved around a fixed sun (as the

Greek philosopher Aristarchus suggested in the third century BCE), and that some or all of the planets revolved around the sun, which in turn revolved around the earth (an idea passed on by Roman writers to the philosophers of the Middle Ages and invented independently in India in the fifteenth century). It was only a thousand years after the fall of Rome, however, that it became generally agreed—and soon after, known for sure—which one of these theories was correct.

That great leap forward was made in the exhilarating period between the years 1600 and 1700, during which empirical inquiry evolved from the freewheeling, speculative frenzy of old into something with powers of discovery on a wholly new level—the knowledge machine. Driving this machine was a regimented process that subjected theories to a pitiless interrogation by observable evidence, raising up some and tearing down others, occasionally changing course or traveling in reverse but making in the long term unmistakable progress. Where Thales once surveyed the horizon and saw water, our radio telescopes look into deep space and see dark matter.

It is to mark this sudden change in the tempo and form of discovery that historians call what happened the "Scientific Revolution," and philosophers and sociologists distinguish what came after the Revolution as a new way of thinking about the world. In so doing, they set "modern science" apart from what preceded it—that is, ancient and medieval science, or what is sometimes called, to emphasize the striking discontinuity, "philosophy of nature" or "natural philosophy." The natural philosophy that came before the Scientific Revolution was not less creative than modern science, and as practiced by thinkers such as Aristotle was no less methodical and no less concerned with the evidence of the senses. Yet something, it seems, was missing.

For that extraordinary something, why the excruciating wait? Why, after philosophy and democracy and mathematics tumbled through the doors of the ancient thinkers' consciousness in quick succession, did science dawdle on the threshold? Why wasn't it the ancient Babylonians putting zero-gravity observatories into orbit around the earth; the Han Chinese building particle accelerators in the flat fields along the Yellow River; the Maya growing genetically modified corn in the Yucatán; the ancient Greeks engineering flu vaccines and transplanting hearts?

Events such as revolutions and elections, declarations and emancipations occur at particular times and in particular places. But a nonevent such as science's non-arrival happens, so to speak, almost everywhere. Modern science did not arrive in democratic Athens. It was not invented by Aristotle. It failed to develop in China a thousand years ago, in spite of that nation's cohesion, scholarly tradition, and technological prowess. Neither Islamic nor European medicine succeeded in becoming truly scientific. The Maya, the Aztecs, the Incas; Goryeo Korea and the Khmer Empire of Cambodia; the India of the Maurya and Mughal Empires: we marvel at their temples and pyramids, their rich traditions of theater and dance. But these cultures, all wealthy, powerful, and sophisticated, are equally non-inventors of science.

The long absence of science is therefore not to be explained by some particular train of events or some specific mix of custom and circumstance. It spans democracies and theocracies, east and west, pantheists and People of the Book. It seems, to all the world, that there is something about the nature of science itself that the human race finds hard to take on board.

That, I believe, is the answer: science is an alien thought form. To understand its late arrival on the human scene, we need to appreciate the inherent strangeness of the scientific method.

The first step is to take a close look at the method, at the rules that drive modern science and explain its fact-finding power. An easy task, you might think. The principles in question regulate science wherever it is found. Every university, every research facility, every industrial laboratory follows them. Go there; make a few queries. Scientists themselves will tell you what science is and how it functions.

Yet it has turned out to be anything but straightforward to obtain a satisfactory answer. Some scientists say that the essence of science is controlled or repeatable experiment, forgetting that experiments are of relatively little importance in cosmology or evolutionary biology. Some say advanced mathematical techniques are crucial, forgetting that the discoverers of genetics, for example, had no use for sophisticated math. Some say what matters is observation. That is a big-hearted answer—it does not exclude whole branches of science like the other responses— but it is too generous. The ancient Greek natural philosophers sought to explain what they saw around them, yet for all their yearning to make sense of the observable world, they had not yet laid their hands on the secret of modern science.

Carry out a poll of the scientific profession, then, and you will soon discover that although scientists know very well how to implement their methods, they don't know what it is about those methods that really matters, and why.

What about other scholars who study the nature of science—the historians of science, the sociologists of science, the philosophers of science? You'll find no more agreement among them than you do among working scientists. Indeed, the question of the scientific method is one of the most difficult, most contentious, most puzzling problems in modern thought.

The consequence is an argument that has sometimes smoldered, sometimes blazed for well over a hundred years—the Great Method Debate. Subtle, powerful thinkers have attempted to describe the scientific method and come to quite opposing conclusions.

More perplexing still, many who have inspected science closely have concluded that there is no such thing as the scientific method. To the question of what is new about modern science, of what changed in the Scientific Revolution, they answer "Nothing much"—or as the sociologist Steven Shapin has declared, "There was no such thing as the Scientific Revolution." Three centuries after Newton explained why the planets orbit the sun, the nature of science is, as the philosopher of science Paul Feyerabend has written, "still shrouded in darkness."

Into that darkness *The Knowledge Machine* will plunge, searching for illumination among the tangle of competing visions of and skepticism about the scientific method. It will wrangle with philosophers such as Karl Popper, who believed that the method hinges on a certain kind of logic applied by thinkers with the right sort of temperament, and Thomas Kuhn, who thought that it is rather a special kind of social organization that is responsible for science's power. It will confront sociologists such as Steven Shapin who hold that no method exists. And it will put forward its own proposal about the nature of the method.

There are many reasons to join the Great Method Debate. Science is so vital to the quality of our modern life that even if the scientific method turned out to be something rather boring and unremarkable—a superior kind of experimental technique, say—it would be imperative to find it and to frame it in a book.

I would have no interest, however, in writing that book. What fascinates me is that science's rules of engagement are so unexpected, so unintuitive, so odd. It is this peculiarity, I believe, that accounts for science's late arrival. Even putting aside the fascinating question of science's delay, the weirdness of the scientific method is an intellectual spectacle in itself. It is to share and delight in that spectacle, as much as anything else, that I put these pages before you.

Once I have taken my turn as the P. T. Barnum of the laboratory, unveiling the monstrosity that lies at the heart of modern science, you will begin to understand why it has been so difficult to chase down.

Those who have searched for a scientific method—the *methodists*—have been looking for a logical and behavioral directive that expunges human whim from scientific thought, replacing it with a standardized rule or procedure for judging theories in the light of evidence that explains science's stupendous knowledge-producing capacity. The rule that governs science and explains its success is far weaker, however, than the methodists have supposed: it tells you what counts as evidence but offers no system for interpreting that evidence. Indeed, it says nothing about the significance of evidence at all.

Further, the rule does not reside where the methodists have expected to find it, in scientists' heads. It does not tell scientists what to think privately; it merely regulates how they argue publicly. It is not a method of reasoning but a kind of speech code, a set of ground rules for debate, compelling scientists to conduct all disputes with reference to empirical evidence alone.

That explains why the methodists failed to find their method. They were looking for the wrong kind of rule. And they were looking in the wrong place.

How can a rule so scant in content and so limited in scope account for science's powers of discovery? It may dictate what gets called evidence, but it makes no attempt to forge agreement among scientists as to what the evidence says. It simply lays down the rule that all arguments must be carried out with reference to empirical evidence and then steps back, relinquishing control. Scientists are free to think almost anything they like about the connection between evidence and theory. But if they are to participate in the scientific enterprise, they must uncover or generate new evidence to argue with. And so they do, with unfettered enthusiasm.

The resulting productivity is what makes all the difference: science is a machine for motivating disputatious humans to carry out tedious measurements and perform costly and time-consuming experiments that they would otherwise not care to undertake. It is these empirical minutiae, so painful to collect, that single out the truth among the plausible falsehoods. Eventually, enough such evidence accumulates that just about every scientist, whatever their quirks, biases, and prejudices, agrees on its significance: one theory stands well above the rest as the best explainer and predictor of it all.

The apparently unassuming code of public behavior, the evidence-only constraint on scientific argument that constitutes all the method science needs to set humanity marching inexorably toward truth, deserves a grand name; I call it the *iron rule of explanation*. Much of *The Knowledge Machine* is given over to understanding where the iron rule came from, what it amounts to, and by what means it leads science toward enlightenment. That will be my attempt to settle the Great Method Debate. If I am correct, then in this book you will discover how science really works.

You will also find the answer to my opening question, learning why it took so long for the human race to discover the power of the iron rule. I won't explain science's late arrival by composing a history of the origins of the Scientific Revolution. My interest is in all the apparently propitious places and times that science *failed* to appear. That nonappearance is to be explained by something timeless: that the iron rule, the key to science's success, is unreasonably closed-minded. It works superbly well, but from the outside, it looks to be, quite simply, an irrational way to inquire into the underlying structure of things. The ancient Greeks had poetry, music, drama, philosophy, democracy, mathematics—each an expression and an elevation of human nature. Science, by contrast, requires of its practitioners the strategic suppression of human nature, indeed, the suppression of the highest element of human nature, the rational mind. What Greek philosopher could have supposed that this was the route to unbounded knowledge of the world? The mystery is not that science arrived so late, but that as a technique for discovery it was ever hit upon at all.

By the end of *The Knowledge Machine*, then, I will have answered two big questions, one philosophical and the other historical:

1. How does science work, and why is it so effective?
2. Why did science arrive so late?

To the first, philosophical question, I say: what matters is the iron rule. To the second, historical question, I say: it is the irrationality of the rule that barred it from human consciousness for so long.

The investigation of the intellectual, moral, and social structure of science that answers both the philosophical and historical questions will constitute by far the greater part of *The Knowledge Machine*, but near the end I will yield to the temptation to do something more like conventional history, explaining why science finally turned up when and where it did, in the European seventeenth century. I then comment on the impact of the iron rule's irrationality on the shape of science today and ask how we can best maintain, even improve, the knowledge machine so as to continue to profit from its penetration and power—and not least, so as to save ourselves from some of the havoc it has helped us to wreak on our planet.

The Knowledge Machine has much to say in favor of science. It sets out to defend scientific inquiry against those who doubt its legendary ability to find the truth—against fundamentalists, postmodernists, romantics, spiritualists, philosophical skeptics. The legend, I show, is firmly founded in fact.

Yet these same arguments and explanations show how peculiar and often inhuman in its workings the knowledge machine can be. It gets the job done not in spite of but in virtue of its proprietary blend of inarticulacy, closed-mindedness, and systematic irrationality. No wonder humanity was so reluctant for so long to pull the lever that brought it buzzing and spluttering to life.

My story begins at the height of the Great Method Debate, as the twentieth century's two foremost philosophers of science—Karl Popper and Thomas Kuhn—lay out opposing visions of the mechanism behind science's knowledge-making capacity. Neither effort will succeed. But sifting through the philosophical wreckage, I will find the foundation on which to construct a better theory of science.

I

THE GREAT
METHOD DEBATE

CHAPTER 1

Unearthing the Scientific Method

ego

Karl Popper's and Thomas Kuhn's profoundly different
theories of the way science works—and the idea
they share that points the way to the truth

I N 1942, Karl Popper was excavating the bones of a gigantic extinct bird, the monumental moa, near Christchurch, New Zealand. A position teaching philosophy just north of Antarctica was his refuge from Hitler's armies, which had marched into his home city of Vienna four years before.

When he was not prospecting for oversized avian remains, he was hard at work writing a condemnation of totalitarianism, in both its Nazi and communist forms, to be published at the end of the war. The twentieth century's political and human chaos showed, Popper thought, that progress of any sort would be possible only through the vigorous exercise of the highest forms of critical thought, and for this Austrian refugee and antipodean adoptee, highest of all was scientific inquiry—perhaps the only human activity, he wrote, "in which errors are systematically criticized and fairly often, in time, corrected." Much of his life's work was therefore given over to an investigation of the rational basis of science, presented to the world in his philosophical masterpiece *The Logic of Scientific Discovery*.

The ideas in that book would change the way generations of physicists, biologists, economists, and philosophers would think about the
scientific method. After Popper moved from New Zealand to take up
an academic post in Britain in 1946, he was elected a Fellow of the
Royal Society, knighted by Queen Elizabeth II, and declared by the
Nobel Prize–winning biologist Sir Peter Medawar to be "incomparably
the greatest philosopher of science that has ever been."

Popper, born in 1902, turned 12 the day that Austria-Hungary
declared war on Serbia, precipitating the Great War. He came of age in
the social and economic devastation that followed. "The war years and
their aftermath," he later wrote, "were in every respect decisive for my
intellectual development. They made me critical of accepted opinions,
especially political opinions." He continued:

> The famine, the hunger riots in Vienna, and the runaway infla-
> tion . . . destroyed the world in which I had grown up. . . . I was
> over sixteen when the war ended, and the revolution incited me
> to stage my own private revolution. I decided to leave school, late
> in 1918, to study on my own. . . . There was little to eat; and as
> for clothing, most of us could afford only discarded army uni-
> forms. . . . We studied; and we discussed politics.

For a few months, Popper threw in his lot with the communists, only
to back away after a violent demonstration led to several protesters'
deaths—a consequence, he thought, not only of police brutality but also
of the demonstrators' tactical aggression.

He remained a socialist, however, and around 1919 resolved to take
up manual work. At this time he was squatting in an abandoned wing
of a former military hospital, feeding himself by tutoring American
university students. The experiment with blue-collar labor turned out
badly: he was, he tells us, too feeble to wield a pickaxe and too distracted

Figure 1.1. Police attempt to contain communist demonstrators in Vienna, June 1919.

by philosophical ideas to produce the straight edges and square corners required of a cabinetmaker. Abandoning these pursuits, he became a social worker, caring for neglected children. Not much later he left behind socialism itself, reasoning that while freedom and equality are both much to be desired, to have both was impossible—and in the end, "freedom is more important than equality."

In the same year, Popper heard Einstein lecture on his new theory of relativity: "I remember only that I was dazed. This thing was quite beyond my understanding." But he was struck by Einstein's willingness to subject his theory to empirical tests that might disprove it:

Thus I arrived, by the end of 1919, at the conclusion that the scientific attitude was the critical attitude, which did not look for verifications but for crucial tests; tests which could *refute* the theory tested, though they could never establish it.

In that single italicized word germinated Popper's greatest and most influential idea.

The idea had its roots in a conundrum posed by the Edinburgh philosopher David Hume in 1739, in the first decades of the Scottish Enlightenment. Imagine Adam, mused Hume, waking in Eden for the first time—naked, alone, wholly unspoiled by knowledge of any sort. Wandering through the primeval woods, he makes some elementary discoveries: fire burns, fruit nourishes, water drowns. Or more exactly, he makes some particular observations: he burns his fingers in some particular fire; he finds some particular pieces of fruit from some particular trees nourishing; he sees some particular animal drown in some particular river. Then he generalizes, using all his newborn wit: best you avoid getting too close to any fire; best to satisfy your hunger by eating fruit from that kind of tree; and so on. This sort of generalization from experience is called inductive reasoning, or, for short, induction.

What, Hume asked, justifies these generalizations? Why is it reasonable to think that merely because this fire burned you yesterday, it will burn you again today? It's not that Hume was recommending that you plunge your hand into the flames any time soon. He just wanted you to explain your reluctance.

There is an obvious answer to Hume's innocent inquiry: things tend to behave the same way at all times—at least most things, most of the time. Fire will tend to affect flesh similarly, yesterday, tomorrow, and next week. So, in the absence of any other information, your best bet for predicting fire's future effect is to generalize from the effects you've already seen. The practice of induction is justified, in other words, by appealing to a universal tendency to regularity or uniformity in the behavior of things. Hume considered this answer, and replied: yes, but what justifies your belief in uniformity? Why think that fire's effects are fixed? Why think that future behavior is in general like past behavior?

There's an obvious answer to that question, too. We think that behavior will be the same in the future as it was in the past because in our experience, it always has been the same. We justify our belief in uniformity, then, by saying that nature has always been uniform in the past, so we expect it to continue to be uniform in the future.

But that, as Hume observed, is itself a kind of inductive thinking, generalizing as it does from past to future. We are using induction to justify induction. Such circular reasoning cannot stand. The snake in the garden swallows its own tail.

There is no other route, Hume thought, by which inductive thinking might be vindicated. He was a philosophical skeptic: he believed that all those inferences that are so vital for our continued existence—what to eat, where to find it, what to pass over—are at bottom without justification. But like many skeptics, he was also a conservative: he advised us to press on with induction in our everyday lives without asking awkward philosophical questions. The English philosopher Bertrand Russell, writing about Hume two hundred years later, could not accept this philosophical quietism: if induction cannot be validated, he wrote, "there is no intellectual difference between sanity and insanity." We will end up like the ancient Greek skeptic who, having fallen into a ditch, declined to climb out because, for all he knew, his future life in the mud would be as good as, perhaps much better than, life above ground. Or as Russell put it, our position won't differ from that of "the lunatic who believes that he is a poached egg."

For all that, there is still no widely accepted justification for induction. Popper saw no alternative but to accept Hume's argument; unlike Hume, however, he concluded that we must abandon inductive thinking altogether. Science, if it is to be a rational enterprise, must not regard the fact that, say, fire has been hot enough to burn human skin in the past as a reason to think that it will be hot enough to burn skin in the future. Or to put it another way, the fact that fire has burned us in the past may not in any way be counted as "evidence for" the hypothesis that fire will be hot enough to burn us in the years to come. Indeed, science ought not to make any use whatsoever of the notion of "evidence for." So there can be no evidence for the hypothesis that the earth orbits the sun (since that implies that the earth will in the future continue to orbit the sun); no evidence for Newton's theory of gravitation; no evidence for the theory of evolution; no evidence in fact for anything that we've ever called a "theory."

This might sound like just the sort of insanity that Russell feared. But Popper was no poached egg. Science, he thought, had a powerful replacement for the inductive thinking undermined by Hume. There may be no such thing as evidence *for* a theory, but what there can be— and here Popper recalled his youthful bedazzlement by Einstein in 1919—is evidence *against* a theory. "If the redshift of spectral lines due to the gravitational potential should not exist," Einstein wrote of a certain phenomenon predicted by his ideas, "then [my] general theory of relativity will be untenable." As Einstein saw, we can know for sure that any theory that makes false predictions is false. To put it another way, a true theory will always make true predictions; false predictions can issue only from falsehood. No assumptions about the uniformity of nature are needed to grasp that.

If your theory says that a comet will reappear in 76 years and it doesn't turn up, there is something wrong with the theory. If it says that things can't travel faster than the speed of light and it turns out that certain particles gaily skip along at far greater speeds, there is something wrong with the theory. And indeed, if your theory says you are a poached egg and you find yourself strolling the London streets on two sturdy legs far from the nearest breakfast establishment, then that theory, too, is wrong. Russell needn't have worried. Unlike inductive thinking, this is all just straightforward, incontrovertible logic.

Such is the logic, according to Popper, that drives the scientific method. Science gathers evidence not to validate theories but to refute them—to rule them out of the running. The job of scientists is to go through the list of all possible theories and to eliminate as many as possible, or, as Popper said, to "falsify" them.

Suppose that you have accumulated much evidence and discarded many theories. Of the theories that remain on the list, it is impossible, according to Popper, to say that one is more likely to be true than any of the others: "Scientific theories, if they are not falsified, forever remain . . . conjectures." No matter how many true predictions a theory has made, you have no more reason to believe it than to believe any of its unfalsified rivals.

Let me repeat that. Popper is sometimes said (by the *New Oxford American Dictionary*, for example) to have claimed that no theory can be proved definitively to be true. But he held a far more radical view than this: he thought that of the theories that have not yet been positively disproved, we have absolutely no reason to believe one rather than another. It is not that even our best theory cannot be definitively proved; it is rather that there is no such thing as a "best theory," only a "surviving theory," and all surviving theories are equal. Thus, in Popper's view, there is no point in trying to gather evidence that supports one surviving theory over the others.

Scientists should consequently devote themselves to reducing the size of the pool of surviving theories by refuting as many ideas as possible. Scientific inquiry is essentially a process of disproof, and scientists are the disprovers, the debunkers, the destroyers. Popper's logic of inquiry requires of its scientific personnel a murderous resolve. Seeing a theory, their first thought must be to understand it and then to liquidate it. Only if scientists throw themselves single-mindedly into the slaughter of every speculation will science progress.

Scientists are creators as well as destroyers: it is important that they explore the theoretical possibilities as thoroughly as they can, that they devise as many theories as they are able. But in a certain sense they create only to destroy: every new theoretical invention will be welcomed into the world by a barrage of experiments devised solely to ensure that its existence as a live option is as short as possible. There can be no favorites. Scientists must take the same attitude to the theories that they themselves concoct as to those of others, doing everything within their power to show that their own contributions to science are without any basis in fact. They are monsters who eat their own brainchildren.

It is carnage, this mass extermination of hypotheses. Yet Popper, the survivor of two world wars, thought it essential to human progress:

> Let our conjectures . . . die in our stead! We may still learn to kill our theories instead of killing each other.

To be an imaginative explorer of new theoretical possibilities and a ruthless critic, determined to uncover falsehood wherever it is found—that is the Popperian ideal. Scientists are both empirical warriors and intuitive artists, combining originality and openness to new ideas with an intellectual honesty that regards nothing as above suspicion.

Tough and tender, hard-eyed yet broad-minded, passionate, courageous, imaginative—who would not sit for such a self-portrait? Working scientists fell head over heels for Popper's ideas. "There is no more to science than its method, and there is no more to its method than Popper has said," proclaimed the cosmologist Hermann Bondi, declaring Popper the uncontested winner of the Great Method Debate. The eminent neuroscientist John Eccles wrote, "I learned from Popper what for me is the essence of scientific investigation—how to be speculative and imaginative in the creation of hypotheses, and then to challenge them with the utmost rigor."

Popperian formulations abound not only in philosophical panegyric but also in practice, most notably in postwar Britain, where Popper made his home. In attempting to undercut the work of the neuroen-docrinologist Geoffrey Harris in 1954, the anatomist Solly Zuckerman declared that a scientific hypothesis "falls to the ground the moment it is proved contrary to any of one of the facts for which it is designed to account"; he then flaunted a single ferret brain that he supposed would annihilate Harris's career.

Popper's contribution to the mythos of science is familiar to many scientists and science lovers. I often wonder whether they grasp, however, how peculiar a view of the logic of science lies at its core—a view on which no amount of evidence can give you more reason to believe a theory than you had when it was first formulated and completely untested; a view on which induction is a lie; a view on which you have no grounds whatsoever to think that the future will resemble the past, that the universe will go on humming the same tune rather than spontaneously changing its song.

Almost every other philosopher of science finds room for induction. Some believe that Hume's problem must have a solution—that is, a philosophical argument showing that it is reasonable to suppose that nature is uniform in certain respects, though we may still be waiting for the thinker clever enough to unravel the Humean knot. Some believe, like Hume himself, that it has no solution but that we must go on thinking inductively regardless, both in our science and in our everyday lives. All believe that induction is essential to human existence. What made Popper different?

Perhaps there is a clue in a story told about Hans Reichenbach, a professor of philosophy in Berlin in the early 1930s. Like Popper, Reichenbach escaped to the English-speaking world as totalitarianism engulfed his Germanic homeland. Reichenbach had not thought much about Hume's worry that the future may fail to resemble the past until 1933. In that year, the Nazis burned the Reichstag, took control of the University of Berlin, and expelled many of its Jewish professors and staff,

Reichenbach included. "Then," Reichenbach is said to have observed, "I understood at last the problem of induction."

REICHENBACH, POPPER, AND many like-minded refugees fleeing the mayhem and malevolence of Central Europe between the wars promoted an ideal of the scientist as a paragon of intellectual honesty, standing up for truth in the face of stifling opposition from the prevailing politics, culture, and ideology.

To this vision, Thomas Kuhn presented the utter antithesis, a dark and deflating conception of the internal machinery of science liable to repel working scientists and on first appraisal quite unsuited to explain science's heroic feats of discovery.

Before becoming a philosopher, Kuhn was a historian of science. Before becoming a historian, he was a physicist. The road was straight and smooth: Kuhn, born in 1922, attended an elite private school in Connecticut and then studied at Harvard for both his undergraduate and his doctoral degrees in physics. His academic career opened with a prize position at the Harvard Society of Fellows, after which he taught at Harvard, Berkeley, Princeton, and MIT. There was no pick swinging or cabinetmaking; he never taught abused youth—except, if the filmmaker Errol Morris is to be believed, his own graduate students. (Morris recalls that Kuhn, a chain-smoker of prodigious capacity, once attempted to refute an objection Morris posed by flinging a loaded ashtray at his head.)

In spite of his early advantages and successes, Kuhn was, he tells us, "a neurotic, insecure young man." He entered psychoanalysis while in graduate school in the 1940s. While he found its therapeutic value to be doubtful, he credited it with enhancing his own interpretive powers to the point that he "could read texts, get inside the heads of the people who wrote them, better than anybody else in the world."

This new ability soon manifested itself in a way that suggested to Kuhn the ideas that would make him famous. Puzzling over Aristotle's theory of physics, which "seemed to me full of egregious errors," Kuhn looked out the window and had an epiphany:

> Suddenly the fragments in my head sorted themselves out in a new way, and fell into place together. My jaw dropped, for all at once Aristotle seemed a very good physicist indeed, but of a sort I'd never dreamed possible. Now I could understand why he had said what he'd said.

Kuhn did not, of course, come to believe Aristotle's physical theory, but he did come to see it as a system that, by its own lights, constituted a coherent and powerful explanatory framework. To appreciate its cogency, however, he had to set aside his habitual ways of thinking about the world, conditioned by twentieth-century physics, and to adopt temporarily a wholly new worldview. From this experience he learned that some revisions of scientific theory are so profound that they require a complete overturning of the cognitive order—a revolution.

Kuhn's famous book *The Structure of Scientific Revolutions* was published in 1962, 15 years after his epiphany and just 3 years after Popper's own great work on the scientific method first appeared in English. Nothing before or since has had a comparable impact on the philosophy of science; nothing has so altered the course of the Great Method Debate. A book on revolutions that took the '60s by storm? You might suppose that Kuhn's picture of science was a model of intellectual ferment, radical thinking, inspired resistance to the choke hold of tradition. Not so. Science is capable of world-altering progress only because, according to Kuhn, scientists are quite incapable of questioning intellectual authority.

Any branch of science—microeconomics, nuclear physics, genetics—

has at all times, says Kuhn, a single dominant ideological mind-set, something he calls a *paradigm*. The paradigm is built around a high-level theory about the way the world works, such as Newton's theory of gravitation or Mendel's laws of genetics, but it contains much more as well: it identifies, in the light of the theory, what problems are important, which methods are valid ways to go about solving the important problems, and what criteria determine that a solution to a problem is legitimate.

A paradigm functions, then, as a more or less complete set of rules and proper behaviors for doing science within a discipline. Scientists obey these rules religiously. To invoke blind devotion is not a metaphor: scientists don't follow the paradigm because they believe it is well supported by the evidence, or because it is the "official" way to do things, or because it is especially well funded, or because it seems like it might be worth a shot; rather, they follow it because they cannot imagine doing science any other way. Were they presented with an alternative paradigm, Kuhn argues, they would find it incomprehensible.

To explain this mental block, Kuhn appealed to experiments in perception conducted by the psychologist Jerome Bruner and others, in which subjects are briefly shown (for example) "trick" playing cards, such as a six of spades printed with red rather than the standard black ink. The subjects report experiencing what their prior beliefs would lead them to expect, rather than what is actually on the card; they might see a black six of spades when what's sitting in front of their eyes is manifestly red, or they might misread the card as a six of hearts. Even the direct evidence of the senses, Bruner concluded, is swayed by our beliefs about what's out there. That's possible, according to Bruner, because our raw experience of things is ambiguous, like the drawing in Figure 1.2. Is it a duck or a rabbit? Apparently a duck . . . but rotate the image a quarter turn clockwise, and it is a rabbit that stares unblinking out of the page. It is our preexisting assumptions, our theories, our prevailing worldview that disambiguate what's supplied by the senses, thereby presenting us with a determinate mental picture of the world.

Scientists, like anyone else, see and understand things at any one time from within a particular worldview. That may sound innocent enough, but it shuts down scientists' capacity to comprehend genuine novelty. To grasp a new worldview, you would need to appreciate it from the perspective of some worldview or other. You can't appreciate it from the new worldview's perspective (that is, its own perspective), because you haven't yet grasped that framework. But if the old worldview is incompatible with the new, then you can't see the new view from the perspective of the old view either. The new view is simply out of sight.

Figure 1.2. Duck or rabbit?

The contrast with Popper is stark. For Popper, what matters above all else to the successful operation of the knowledge machine is scientists' acute faculty for critical thought. They can survey the theoretical possibilities, and they see clearly how each theory might, in the face of the evidence, collapse. For Kuhn, such a survey, the essential precondition for criticism, is psychologically impossible.

In supposing that scientists could not simultaneously contemplate rival grand theories, Kuhn was putting enormous conceptual weight on a few empirical findings and philosophical arguments, no doubt inspired by his own experience with Aristotle's physics. He was moving with the zeitgeist, however, and his readers, or enough of them, went along with it. When Kuhn's book appeared in 1962, it was still the age of the military-industrial complex, the man in the gray flannel suit, and William Whyte's "organization man"—a complacent and compliant figure

Figure 1.3. Organization men.

eager to fit into the system and carry out whatever plan was handed down from above.

The prevailing paradigm's staffers cannot conceive of any other way to do science. And yet, Kuhn observes, a paradigm is not forever. Existing ideas crumble during events that historians call scientific revolutions, intellectual cataclysms during which a new paradigm replaces the old. (A lowercase scientific revolution should not be confused with the uppercase Scientific Revolution, of which there has been only one. In a lowercase scientific revolution, one way of doing science is replaced by another. In the uppercase Scientific Revolution, something that was not science—natural philosophy, I have called it—was replaced by a far more effective form of empirical inquiry, modern science itself.)

Before Kuhn wrote about scientific paradigms, he wrote a history of the sixteenth- and early seventeenth-century Copernican revolution, arguably the first scientific revolution of all. The old regime, overthrown by the revolution, was the ancient Greek system of astronomy, perfected in the work of the Greco-Egyptian mathematician Ptolemy, according to whom the sun, the moon, the stars, and all the planets orbit the earth. The revolutionary new idea was Copernicus's system, published in 1543, in which the moon orbits the earth and everything else, including the earth, orbits the sun. As developed by Johannes Kepler in the early 1600s, it predicted the paths of the celestial bodies more accurately and more elegantly than Ptolemy's theory.

A shift to the Copernican system, in spite of its predictive superiority, was at the very least troubling. It meant taking on board the rather deflating realization that the earth is not, after all, the center of the universe—though a certain grim satisfaction could perhaps be had from the accompanying realization that there is no distinction between the corrupt earth and the supposedly perfect, symmetrical, unblemished heavens, that every celestial body is equally rough-hewn, dog-eared, moth-eaten, coarse.

A less soulful and more visceral drawback of Copernicanism was its implication that the earth is moving very fast—rotating every 24 hours and racing around the sun in 365 days (at a speed, we now know, of about 66,600 miles per hour). How could we not have noticed? The answer lay in a second and parallel revolution in physics that accompanied the revolution in astronomy. The radical new physical idea was that a person or thing moving at an approximately constant velocity, like the seas and trees and people on the surface of the earth, will not experience the speed at all; however fast they're moving, they will feel as though they're standing still.

It was not easy for human minds to let go of the centrality of the earth, the perfection of the heavens, and the palpability of speed. This intellectual stasis Kuhn put down to the paradigm's stifling embrace. Copernicus triumphed all the same. And from then on, paradigms continued to topple. Newton's theory of gravity replaced Aristotle's story, according to which rocks fall to earth because they are seeking their proper place at the center of the universe, along with various notions of the medieval philosophers. In the nineteenth century, Darwin's theory of evolution by natural selection replaced the theory of special creation, according to which every species was created separately by God. And shortly after the beginning of the twentieth century, Newton's physics was replaced in turn by Einstein's theory of relativity and by quantum physics.

How does this happen? How do paradigms end? A scientist working within a paradigm is not seeking to undermine it. On the contrary, according to Kuhn, they have no inkling that it can be undermined, or at least they don't regard its being overthrown as a serious possibility: "Normal science . . . is predicated on the assumption that the scientific community knows what the world is like. . . . [It] does not aim at novelties of fact or theory and, when successful, finds none." But scientists' very commitment to the paradigm can push it to the point of destruction: they abide by its prescriptions, they faithfully execute its plan, yet they run into insoluble problems because the paradigm is inadequate in some way. From on high, the paradigm guarantees that a certain method will result in answers; following the method, however, leads increasingly to questions, problems, inconsistencies, perplexities. Planets stray from their assigned paths; fossils are unearthed suggesting that human ancestors bore a startling resemblance to apes; light itself can't decide whether to act as a particle or as a wave. The result is what Kuhn calls crisis, a progressive decline of researchers' faith in the paradigm's power.

Without faith, a Kuhnian scientist is lost. The only recipe they have for doing science is the one prescribed by the paradigm that looks to have deserted them. Their enthusiasm for the old system of belief is gone, but if they are to be a scientist at all, they must follow its rituals nevertheless.

There things might hang for decades or longer. Eventually, however, some visionary "deeply immersed in crisis" is able to shrug off the pull of the old ideas; a new way of doing things comes to them "all at once, sometimes in the middle of the night." The prevailing paradigm has competition at last. Given its inadequacies, scientists ought to grasp hungrily at any promising alternative. So they would, perhaps, if they knew what they were grasping for. On Kuhn's understanding of the scientific mindset, however, it is impossible for an adherent of one paradigm to appreciate or even to understand the significance of another. (Kuhn writes that the creators of new paradigms escape the pull of the old because they are "either very young or very new to the field"; their minds have yet to set.)

Here is the predicament, then, of scientists who grew up with the old paradigm—such as adherents of Ptolemy when the Copernican revolution crested in the seventeenth century, or of Newton as Einstein precipitated the twentieth-century revolution in gravitational theory. They know that something has gone badly wrong. Their paradigm has ceased to bestow scientific blessings. Weariness and confusion have taken hold. At the same time, they know there is a new paradigm. They don't themselves understand it, but they see that its followers have all the enthusiasm and joy in discovery that has trickled away from their own intellectual lives. What to do?

Some adherents of the old paradigm will die disillusioned. Some will fight theoretical novelty to the end. But some, the apostates, will undertake to abandon the old theory and to make a move to the new. They will set out to live among its followers or, if that is impossible, to immerse themselves in the new paradigm's canonical writings. Eventually, if conditions in the minds of these apostates are right, the new doctrines will come to supplant the old. The scientist will have undergone what Kuhn calls a "conversion experience."

If the new paradigm is sufficiently fruitful, and its followers dedicated enough in their scientific missionary work, almost every remaining adherent of the old paradigm will, feeling their life's former foundation sinking under their feet, throw themselves into the new way of doing things, the new theory. A scientific revolution will have occurred.

Kuhn scandalized the world of science with this picture of revolutionary scientific change. Previous historians and philosophers had seen scientific change as a largely rational process: the ideas of Copernicus, of Kepler, of Galileo, of Newton, however radical, were accepted because they were so clearly superior to the old ideas, both in their predictive successes and in their explanatory beauty.

If Kuhn is right, then this older, more dignified conception of scientific progress must be wrong, for in Kuhn's view, it is impossible to compare paradigms: "When paradigms enter, as they must, into a debate about paradigm choice, their role is necessarily circular." Perhaps if you had two brains as you have two hands, you could weigh one paradigm against another. But you have only a single brain, and a single brain is capable of grasping only a single paradigm. You cannot simultaneously appreciate the merits of the Aristotelian and the Newtonian worldviews any more than you can simultaneously be a fervent Muslim and a devoted Roman Catholic. At the height of his rhetoric, Kuhn wrote that the Aristotelian and the Newtonian live in different worlds; you can live in one world or the other, but you cannot be in two different places at the same time. A rational comparison of competing paradigms is therefore humanly impossible.

In the place of logical evaluation, Kuhn posits a leap of faith: a giddy jump through ideologically empty space from the traditional view of things to the revolutionary way of thinking, undertaken in the hope that life will somehow be better under a new scientific sign.

You might imagine what Popper, quitting the Old World with open-eyed defiance, would say about this blind lunge into theoretical darkness, what he would think about Kuhn's contention that "as in political revolutions, so in paradigm choice—there is no standard higher than the assent of the relevant community." Popper's student Imre Lakatos, also a refugee from European totalitarianism, accused Kuhn of making science a matter of "mob psychology."

Figure 1.4. To grasp many paradigms takes many minds. Portrait of Thomas Kuhn by Bill Pierce for *Life* magazine.

Kuhn's critics were sickened by the thought that the major transitions in scientific thinking were episodes of conversion rather than careful deliberation. But equally, they were puzzled by Kuhn's faith that an arational process could have led us to the state of scientific sophistication we enjoy today. If it is impossible to compare objectively the merits of Ptolemaic and Copernican astronomy, how did we get the structure of the solar system right? How did we figure out that the earth does indeed go around the sun, rather than vice versa?

Some of Kuhn's radical followers insinuated that we believe our paradigm is a great improvement over earlier ideas for the same reason that we believe our religion is true or our child is beautiful—not because the empirical evidence says so, but because it is *ours*. Kuhn himself, at least in his later writings, repudiated this view and argued for real progress in science. The Copernican paradigm genuinely is better, objectively, than the Ptolemaic paradigm, he held, because it has superior "puzzle-solving ability." One kind of puzzle is the problem of predicting the future; the theories of Copernicus and Ptolemy both aspire, for example, to forecast the paths of the planets across the night sky. A part of what Kuhn is saying, then, is that later paradigms tend to have more predictive power than earlier paradigms. This growth in predictive power, and not something more parochial, is what accounts for our sense that scientific knowledge is seeing ever more deeply into the nature of things, along with our ability to perform ever more impressive feats with that knowledge—to talk across continents, to fly around the globe, to walk on the moon.

The later Kuhn believed, then, that when scientists make the jump from an old to a new paradigm, they tend to jump from a less to a more predictive paradigm, though they are incapable, as they launch themselves, of appreciating the underlying reasons for the new paradigm's superior future-predicting potential. This restores a pinch of rationality to scientific proceedings: Kuhn's revolutionaries are making covert cost-benefit calculations even as they surge through the streets, subtly targeting their leaps of faith in the general direction of predictive and other kinds of puzzle-solving power.

THE KUHNIAN SCIENTIST IS, when not in revolt, a pedestrian character, dull and deferential. But science itself, Kuhn believed, is supreme among belief systems in its ability to create new knowledge. It is far from the only thought system capable of generating novel and original ideas—philosophy, for example, is its equal in this respect. What is unparalleled is its ability to test those ideas thoroughly, to drive them to their logical or illogical conclusions. Central to science's extraordinary rigor is precisely the limitedness of the individual scientist, their inability to see outside the prevailing paradigm. This intellectual blindness is, then, the core of Kuhn's answer to my big philosophical question about science, the question of what arrived in the Scientific Revolution that made scientific inquiry so much more fruitful than the natural philosophy that had come before.

That science's success is explained by a kind of intellectual confinement—that is the single most astonishing thesis in Kuhn's celebrated book. It is easy to see how the characteristic intellectual demeanor of the Popperian scientist—unbounded imaginer, unrelenting refuter—might sustain the extraordinary productivity of the knowledge machine. But Kuhn's scientists? How could their inability to contemplate or even comprehend new ideas possibly drive discovery?

Science is boring. Science is frustrating. Or at least, that is true 99 percent of the time. Readers of popular science see the 1 percent: the intriguing phenomena, the provocative theories, the dramatic experimental refutations or verifications. Behind these achievements, however—as every working scientist knows—are long hours, days, months of tedious laboratory labor. The single greatest obstacle to successful science is the difficulty of persuading brilliant minds to give up the intellectual pleasures of continual speculation and debate, theorizing and arguing, and to turn instead to a life consisting almost entirely of the production of experimental data.

Many important scientific studies have required of their practitioners a degree of single-mindedness that is quite inhuman. Through the 1960s, the rival endocrinologists Roger Guillemin and Andrew Schally fought to be the first to find the structure of the hormone TRH, a substance used by the hypothalamus, a small but crucial structure at the base of the brain, to set off a chain of signals controlling processes ranging from daily metabolism to early brain development. The full significance of TRH is not yet understood, but some sense of its importance and power can be discerned from the US Army's commissioning, in 2012, of a study to examine its possible use in a nasal spray to quell suicidal urges.

Guillemin and Schally finished in a dead heat, sharing the 1977 Nobel Prize in Physiology or Medicine for their discovery of TRH's molecular makeup. It had been not so much a race as an epic slog. Literally tons of brain tissue, obtained from sheep or pigs, had to be mashed up and processed to obtain just 1 milligram of TRH for analysis. Several rivals dropped out of the competition, unable to countenance the "immense amount of hard, dull, costly, and repetitive work" required. As Schally later explained:

> Nobody before had to process millions of hypothalami. . . . The key factor is not the money, it's the will . . . the brutal force of putting in 60 hours a week for a year to get one million fragments.

Still, the investigation of TRH was over in a flash compared with the Gravity Probe B experiment at Stanford University, which undertook to launch a satellite into orbit around the earth that would measure the "geodetic" and "frame-dragging" effects implied by Einstein's general theory of relativity. The project was initiated in 1964 and made its final report to NASA—after overcoming extraordinary setbacks and technical problems and creating, as components of its gyroscopes, the

Figure 1.5. The rotors for the gyroscopes in the Gravity Probe B experiment—"the roundest objects ever manufactured." They are 1.5 inches across.

most perfectly spherical objects ever fashioned by human hands—in 2008 (Figure 1.5). The director of the project, Francis Everitt, stuck with it for all four-plus decades, 74 years old when he signed that report.

In another 40-year epic, the evolutionary biologists Peter and Rosemary Grant have since 1973 spent their summers on the tiny Galápagos island of Daphne Major, observing, trapping, numbering, and measuring finches in order to demonstrate "evolution in action" as body and beak size adapt over generations to drought, flood, and other environmental changes (Figure 1.6). In 1981, they began to track, in particular, a finch that was larger and had a different song than any known variety. Thirty-one years later, having followed that finch's offspring bird by bird for six generations, they had enough data to conclude that they had observed the origin and establishment of a new species.

A longitudinal study in economics or medicine can likewise involve decades of data collection: the Dunedin Multidisciplinary Health and Development Study has been monitoring a thousand New Zealanders since the early 1970s and will continue into the 2020s.

Figure 1.6. The Galápagos island of Daphne Major is neither large nor hospitable. It is less than half a mile across.

Such Herculean efforts would perhaps be worth countenancing if the data generated by the experiments were guaranteed to result in major theoretical revelations. But as Kuhn noted, the relevance of experimental inquiry frequently hinges on the validity of the paradigm: if there is something conceptually or factually wrong with the method, the results may be of negligible importance.

To detect the frame-dragging effect, the Gravity Probe B apparatus needed to find a rotational change in its gyroscopes on the order of one hundred-thousandth of a degree per year, that is, a change that would take 36 million years to turn the gyroscope rotor in a full circle. Such a microscopic movement could have scientific significance only given a raft of specific physical assumptions. Were any of those assumptions false, the probe's painstakingly precise, excruciatingly expensive measurements would be worthless.

By the time the Dunedin study in New Zealand concludes, plenty more will have been learned about human health from other sources. The project labors, then, in the shadow of the possibility that information about some previously unknown crucial variable is being inadvertently neglected or that some variable thought to be crucial is unimportant—as was the case in the first longitudinal study ever conducted, Lewis Terman's decades-long "Genetic Studies of Genius," which assumed a tight correlation between IQ and genius that decades later turned out not to exist. And the Grants' meticulous finch counting might not have uncovered any particularly interesting patterns of population change, let alone the appearance of a new species; their hard labor and privation would in that case have been for the sake of nothing much at all.

The same is true for scientific investigation on a more modest scale. In a typical physics experiment, it may take years simply to get the apparatus to function properly; in cognitive psychology or the life sciences, it may take years to run pilot studies and to rehearse experimental designs seeking something that will deliver a significant outcome.

The geochemist and biologist Hope Jahren tells of a summer she spent in Colorado monitoring the flowering of a group of hackberry trees. Her aim, part of her PhD research at Berkeley, was to determine the effect of temperature and water chemistry on the composition of the hackberries' fruit. The trees never bloomed; there was no fruit. Jahren's summer was wasted. She asked a phlegmatic local why there were no flowers. The answer? "It just happens sometimes." So she got in her car and drove back to California.

Even when the machinery is running smoothly and the statistics flow plentifully, the results characteristically concern some abstruse matter—the structure of a plant's seed case; the time taken to react to a contrived visual stimulus; the pattern of bright and dark created by intersecting beams of light—whose value rests entirely on the significance it accrues within a larger theoretical framework. What if that framework is mistaken? Years of work, years of life, wasted on the minute inspection of inconsequential trivia.

Science has, as a consequence, a problem of motivation. It is not the problem of motivating students to become scientists; that they might do for many reasons, not least the thrill of discovery. Nor is it the problem of motivating scientists to turn up to the lab each day—they get paid for that—or to observe, measure, experiment when they get there, since that is a standard part of the job description. It is the problem of motivating the extraordinary intensity and long-term commitment with which empirical testing must be carried out in order to do the most valuable science.

How to persuade scientists to pursue a single experiment relentlessly, to the last measurable digit, when that digit might be quite meaningless? "You have to believe that whatever you're working on right now is *the* solution to give you the energy and passion you need to work," says the MIT physicist Seth Lloyd. Or as Andrew Schally wrote about his search for the structure of TRH and other molecules:

Only a person such as myself with strong faith in the presence of these materials would have the patience to go through the many fastidious steps of the isolation procedure.

That is the Kuhnian answer to the motivation problem: mold scientists' minds so that they fail to see that their research might be based on an error, on a false presupposition. If the validity of the paradigm is accepted without question, then the value of long and arduous empirical toil is also beyond question. The purpose of narrowing scientists' horizons is to encourage them to work harder, to dig deeper, to go further than they would go if they could see their destination in perspective, if they had an accurate sense of their project's proportion.

Ultimately, it is only because scientists' faith in the paradigm guarantees the importance of their research that they feel secure enough to work the paradigm to death—to experiment in such detail, with such precision, as to expose the paradigm's shortcomings, to drive science to a crisis, and so to establish the preconditions for revolution. This is Kuhn's marvelous paradox: *A paradigm can change only because the scientists working within it cannot imagine it changing.* It is their certainty of its success that secures its destruction.

POPPER AND KUHN, though different in so many ways, were equally right about some exceptionally important things.

First, they were correct in thinking that what is special about science—what distinguishes scientific thought from the philosophical thought that preceded it—is not so much the capacity to generate new theories as the capacity to eliminate old theories, removing them permanently from humankind's running list of viable options. In either philosopher's story, science's success is due to the unbending search for and pitiless exploitation of even the most minute discrepancies between theory and evidence.

Second, Popper and Kuhn were right in thinking that in order to explain science's critical power, proprietary forms of motivation are at least as important as proprietary logical tools. The tools tell you what to do with the evidence, but that is of no use unless you have the right kind of data, and plenty of it. The production of such data requires, in most cases, an intense and prolonged focus on details of little intrinsic interest. Scientific inquiry needs something, then, to induce thinkers to devote their lives to an enterprise that is in its daily routine mundane and largely negative—while discouraging them from the glamorous alternative, the philosophical strategy of inventing new ideas and new styles of thought at every turn.

Popper finds his motivation in the immense appetite for refutation shared by every good scientist. Kuhn's motivator is more subtle and a little sinister. Individual Kuhnian scientists are not critics at all; they accept the prevailing paradigm with barely a contrary thought. But in their enthusiasm to squeeze every last drop of predictive power from that paradigm, they crush the life out of it.

Science's empirical implacability is, for both Popper and Kuhn, possible only because scientists adhere scrupulously to a method. For Popper that method is universal, fixed for all time—falsification is *the* scientific method. For Kuhn, the method is prescribed by the paradigm, and so it changes whenever scientific revolutionaries impose a new recipe for doing research. The beauty of the Kuhnian story is that it doesn't much matter what the recipe is, provided that it is sensitive to puzzle-solving power, and in particular, to predictive power. Even as the method itself mutates, the fact that science is method-bound, paradigm governed, endows it with its falsifying power. Kuhn is therefore, like Popper, what I have called a "methodist": a believer in the importance of scientists' dutifully following a set formula for pursuing their theoretical inquiries.

The method matters because it exposes predictive deficiencies, but also because it gives scientists the confidence to press on with their experimental lives. Popperian scientists know that since the logic of fal-

sification is indisputable, their colleagues will attribute the same significance to their experimental labors that they do themselves. Kuhnian
scientists have the same expectation because they know that their colleagues subscribe to a single set of rules inherent in the governing paradigm. It is not enough that the rules make sense, then. They must be
widely agreed to make sense. On this matter, too, I think that Popper
and Kuhn are correct.

The Knowledge Machine will build its own explanation of science's
success on these insights, these contributions by Popper and Kuhn to the
Great Method Debate. But I must first explain why modern theorists of
science almost universally reject both thinkers' ideas.

Popper's and Kuhn's theories are not merely philosophical; they
make claims about the actual organization of science and about the
way the organization changes over time. To assess the theories, then, it
makes sense to turn to specialists in these matters, namely, sociologists
and historians of science.

Does contemporary science display the paradigmatic structure
described by Kuhn, in which a single ideology and methodology guides
all scientists working in any given domain? Ask the sociologists. Was
there a sudden and unprecedented onset of paradigm-governed group-
think in the Scientific Revolution? Ask the historians. Do scientists
fight to preserve the status quo, as Kuhn's theory would tend to suggest,
or to overthrow it, as Popper would have it? For contemporary scientists,
ask the sociologists; for the scientists of yore, the historians.

Over the past few decades, the answers have come in. They are
almost entirely negative. There is little evidence, as you will see, for a
dispassionate Popperian critical spirit, but also little evidence for universal subservience to a paradigm. Indeed, in their thinking about the
connection between theory and data, scientists seem scarcely to follow
any rules at all.

CHAPTER 2

Human Frailty

ℰℋℐ

*Scientists are too contentious and too morally and intellectually
fragile to follow any method consistently.*

A s the moon's disk crept across the face of the sun on May 29,
1919, a new science of gravity hung in the balance. Just a few
years earlier, Albert Einstein had formulated his theory of general
relativity, a conceptually radical replacement for the gravitational theory that made Isaac Newton famous at the beginning of modern science,
more than two hundred years before. Whereas Newton held that massive
bodies exert upon each other a "force of gravity," Einstein said that they
rather bend the space and time around them, giving it a characteristic
curvature. When objects do their best to trace straight lines through this
twisted medium, they move in a way that suggests the existence of gravitational force—but there is in fact no such thing. Profoundly different
though these two pictures may be, they make nearly identical predictions
about the movements of particles, planets, and everything in between.
Nearly identical, but not quite. The difference between Newton and Einstein, the fact as to whose ideas were correct, could perhaps be faintly
discerned on the margins of a total eclipse of the sun.

Two months earlier, the steamer *Anselm* had left Liverpool with three telescopes and two teams of scientists on board. One group was headed to Brazil; the other to the island of Príncipe, off the coast of West Africa. At their assigned destinations, they would each photograph the sky at the moment that the light of the sun was fully obscured by the moon. The pattern of stars surrounding the eclipse would reveal the extent to which light passing close to the sun is dragged off course by our home star's intense gravitational field. In the same way that a partially submerged oar appears to bend at the point where it enters the water, due to the bending of light rays at the air/water boundary, so the stars would appear to be displaced from their usual positions to a degree corresponding to the bending force of the sun's gravity. Einstein's new theory predicted that incoming light rays would be deflected by twice the amount that Newton's old theory implied.

It was a crucial experiment in the Popperian mold. Measure the apparent shift in the stars' positions, and in the cold light of that number at most one theory could survive—either Einstein's or Newton's—or, if both predictions turned out to be wrong, neither.

Six months after the eclipse, the expedition leader Arthur Eddington announced the results: Newton was dethroned and Einstein was declared the new emperor of gravitation. The Great War was finally over, and Einstein's esoteric German physics had been confirmed by Eddington's exacting British experiment, a scientific triumph that was heard around the world—by a young Karl Popper among others—and that heralded an era of international cooperation, progress, and peace.

But the peace didn't last, nor did the story; nothing about it is quite right. Eddington awoke on the morning of the eclipse to cloudy skies over Príncipe; he was able to obtain only blurry, indistinct photographs of the surrounding stars. The stellar snapshots from Brazil were much superior, but they posed a different problem. The Brazilian team had brought with them two telescopes, and the measurements made with those telescopes said two different things. One instrument, the "4-inch"

Figure 2.1. Another cloudy day on Príncipe.

telescope, showed a shift in the positions of the stars roughly in accordance with Einstein's prediction. But the other, the "astrographic" telescope (especially designed for photographing stars), showed a shift that was almost exactly Newtonian.

How did Eddington and his collaborators reach the conclusion, then, that it was Einstein's predictions that came true?

They had three sets of data at their fingertips. First, there were 2 photographs from Príncipe that dimly depicted stars through the clouds, and which according to some rather complex calculations by Eddington showed a shift of Einsteinian magnitude. Second, there were 7 photographs from the Brazilian 4-inch telescope that also showed an Einsteinian shift (Figure 2.2 among them). Third, there were 18 photographs from the Brazilian astrographic telescope that showed the shift predicted by Newton's theory. Eddington's strategy was to argue that something had gone systematically wrong with this last set of photo-

Figure 2.2. A photographic plate from Eddington's 1919 eclipse expedition. It is a negative: the eclipsed sun is the big white circle, its bright corona the dark flare surrounding the circle, and the surrounding stars are tiny black dots. Some of the crucial stars' positions are indicated by faint horizontal lines drawn on the plate.

graphs. They were, in fact, considerably blurrier than those produced by the 4-inch telescope, possibly (so he and his collaborators conjectured) because of distortions caused by the sun's uneven heating of the mirror that reflected the light from the eclipse into the telescope.

Certain of Eddington's contemporaries, however, found Eddington's argument to be rather fishy, as have many later historians of science. Eddington could explain the blurriness of the astrographic photos, but he gave no reason to think that they would systematically err so as to give Newtonian rather than Einsteinian values for light's degree of gravitational bending. Further, the crisp photos from the 4-inch telescope gave a value for gravitational bending that was considerably greater than that predicted by Einstein, to a degree that they could be considered to support Einstein's theory only if that telescope, too, was assumed to be systematically biased. Eddington appeared to be engaged in some rather special pleading, then: he assumed systematic errors in one direction for one of the Brazilian telescopes and in the other direction for the other telescope, so as to reach the conclusion that the results they delivered were quite consistent with Einstein's theory of relativity. As W. W. Campbell, an American astronomer and director of the Lick Observatory in San Jose, California, wrote about Eddington's analysis in 1923: "the logic of the situation does not seem entirely clear."

If Eddington's reasoning was as murky as his Príncipe photographs, his aim was pellucid. He wanted very much for Einstein's theory to be true, both because of its profound mathematical beauty and because of his ardent internationalist desire to dissolve the rancor that had some Britons calling for a postwar boycott of German science. (Eddington, as a Quaker, was a committed pacifist; protesting against the proposed boycott, he wrote that "the pursuit of truth . . . is a bond transcending human differences.") These high-minded goals he pursued using the considerable political power at his disposal. He had recruited the Astronomer Royal, Sir Frank Dyson—"the most influential figure in British astronomy"—to his cause early on; it was Dyson who, though he had no personal interest in relativity theory, proposed the eclipse expedition and then took the honorary position as principal author of the expedition's report, all at Eddington's behest.

When the expedition presented its results, Eddington won an endorsement from the president of the Royal Society and qualified support from the president of the Royal Astronomical Society. Other physicists were more dubious but also less influential and less institutionally powerful. Their reservations were written out of the story: in the wake of the eclipse, Eddington became the preeminent exponent of relativity theory in English, and his discussion of the eclipse experiments was regarded as the standard reference on the topic. While he details and celebrates the pro-Einsteinian measurements provided by the Brazilian 4-inch and Príncipe telescopes, the Brazilian astrographic results that favored Newton instead are perfunctorily dismissed. Those photographs were on the wrong side of history; consequently, they were entirely blotted out.

I have begun this chapter with the story of Eddington and the eclipse in part because there is nothing remarkable about it: it is a rather typical (if unusually well documented) tale of complicated, confused, or ambiguous data, a certain selectivity in the interpretation or reporting of that data, and a concerted effort after the report is made to bend the course of consensus making in a direction favorable to the reporter's intellectual, moral, or practical aspirations. This is the human mind operating according to its standard specifications, following a trajectory familiar to every student of history—a pattern of partiality and politicking found in Thucydides's description of the war between ancient Athens and Sparta, in Gibbon's *Decline and Fall of the Roman Empire*, in the intrigues of Renaissance Italy's city-states, and in backrooms and presidential palaces across the globe today.

But scientific reasoning is supposed to be an antidote to these primeval inclinations—and that is what is supposed to explain its extraordinary success. According to Karl Popper, the scientific knowledge machine is driven by an intense critical spirit and by the implacable principle of falsification. Neither is at all evident in Eddington's treatment of the eclipse. Eddington cosseted his own favored theory, shielding it from evidence that looked prima facie falsifying, while damning its rival using reasoning more redolent of the one-sided pleading of a criminal prosecutor than of the evenhanded and straightforward logic of falsification.

According to Thomas Kuhn, what distinguishes scientific from ordinary inquiry is scientists' agreement to conduct their research in the framework of the prevailing paradigm, which both sets their goals and instructs them in the interpretation of the evidence. But there is little sign of such a rigid scaffolding in the case of the eclipse. Eddington used his scientific work to realize an aim that lay outside anything that might be dictated by a Kuhnian paradigm, namely, a rapprochement between the British and the German scientific establishments. Further, he pursued this and his other aims by interpreting the data in a way that seems driven more by the desire to succeed than by some officially sanctioned, widely accepted procedure for bringing evidence to bear on theory, of the sort that a paradigm is supposed to prescribe. His subsequent political machinations and selective history writing equally seem more inspired by personal, albeit idealistic, ambition than by obeisance to a shared code of scientific conduct.

Science is so exceptionally powerful, Kuhn argued, because the supremacy of the paradigm guarantees to scientists (so they believe) that their research has a certain fixed significance, underwritten by the goals, experimental methods, and rules for evaluating evidence that constitute the paradigm's core. Eddington's logical and political manipulations, however, disclose exactly the kind of flexibility of rule and pliancy of institutional framework that would set the significance of scientific results perpetually adrift. The Kuhnian paradigm is supposed to preclude such inconstancy. It did not.

The 1919 eclipse is only a single example of the selective use of evidence. But the centuries since the Scientific Revolution are strewn with cases in which science's biggest names can be seen discarding or distorting difficult data so as to create the impression that experiment was in perfect harmony with their theoretical or other aims.

Gregor Mendel, the founder of genetics, almost certainly massaged the statistics he presented in the 1860s in support of his thesis that genes lie at the root of biological inheritance. Ernst Haeckel embellished his careful drawings of animal embryos around the same time to support his thesis that "ontogeny recapitulates phylogeny"—that a human embryo, for example, passes through stages in which it takes on forms more or less identical to those of fish embryos, then amphibian embryos, then bird embryos. Robert Millikan, in pulling together the data from which he inferred the electric charge of a single electron—work that earned him the 1923 Nobel Prize in Physics—omitted many measurements that did not "look right," while claiming to have included everything. Even Isaac Newton manipulated certain empirical quantities to better fit his theories, tactics that in one case amounted, wrote his biographer Richard Westfall, to "nothing short of deliberate fraud."

There is one respect, I must note, in which Eddington and other modern scientists are almost exceptionlessly careful and methodical. In Eddington's original presentation of the eclipse experiments, you will find certain rules of reporting scrupulously followed. Let your eyes surf over the two tables from Eddington's report reproduced in Figure 2.3. That's scientific method made palpable. There's nothing feigned or dishonest about it. In the upper table is a careful accounting of each of the 18 photographic plates taken with the Brazilian astrographic telescope: the time and length of the exposure and the type of plate are noted. In the lower table are results calculated from the apparent positions of the stars in these plates (omitting plates that showed an insufficient number of stars). The numbers that matter most are in the right-hand column: these give the value of gravitational light bending suggested by each plate. At the bottom right-hand corner is the average of these values, which in a single number summarizes what all the photographs taken using the Brazilian astrographic telescope have to say about gravity's effect on light. That "astrographic bending number" is 0.86, almost exactly equal to the Newtonian prediction of 0.87 and less than half the Einsteinian prediction of 1.74.

If the systematicity and objectivity of science can be seen in the painstaking measurement and calculation and the transparent presentation of the astrographic bending number, the subjectivity and unruliness of science can be seen in what happened next: the number, with its Newtonian implications, was brushed off in a few sentences by Eddington, declared unimportant by his allies in the British scientific establishment,

EXPOSURES with the 13-inch Astrographic Telescope stopped to 8 inches.

Ref. No.	G.M.T. at Commencement of Exposure.				Exposure.	Plate.	Ref. No.	G.M.T. at Commencement of Exposure.				Exposure.	Plate.
	d.	h.	m.	s.	s.			d.	h.	m.	s.	s.	
1	28	23	58	23	5	O.	11	29	0	1	7	5	S.R.
2				37	10	E.	12				22	10	E.
3				57	5	E.	13				36	5	E.
4			59	11	10	S.	14				51	10	S.R.
5				30	5	S.R.	15			2	10	5	S.R.
6				45	10	S.R.	16				25	10	S.R.
7	29	0	0	4	5	S.R.	17				44	5	E.
8				19	10	E.	18				58	10	E.
9				39	5	E.	19			3	18	5	O.
10				53	10	S.R.							

No. of Eclipse Plate.	Ref. No. of Comparison Plate.	No. of Stars.	Values of d, e, α in Revolutions at 50′ Distance.			α at Sun's Limb in Arc.
			$d.$	$e.$	$\alpha.$	
			r	r	r	$''$
1	18_4	7	$+0·051$	$+0·089$	$+0·033$	$+1·28$
2	18_4	11	$-0·009$	$+0·059$	$+0·025$	$+0·97$
3	18_4	8	$-0·074$	$+0·101$	$+0·028$	$+1·09$
4	18_4	11	$-0·168$	$+0·091$	$+0·033$	$+1·28$
5	11_3	10	$+0·094$	$+0·076$	$+0·025$	$+0·97$
6	11_3	11	$+0·186$	$+0·082$	$+0·021$	$+0·82$
{ 7	14_3	12	$+0·006$	$+0·119$	$0·000$	$0·00$
7	18_3	7	$-0·054$	$+0·166$	$0·000$	$0·00$
8	14_3	10	$+0·093$	$+0·064$	$+0·021$	$+0·82$
9	17_4	7	$-0·096$	$+0·129$	$+0·008$	$+0·31$
10	17_4	10	$+0·090$	$+0·045$	$+0·026$	$+1·01$
11	11_1	10	$+0·073$	$+0·061$	$+0·032$	$+1·24$
{ 12	11_1	11	$-0·009$	$+0·102$	$+0·049$	$+1·91$
12	17_2	7	$-0·102$	$+0·114$	$+0·019$	$+0·74$
15	15_3	6	$+0·111$	$+0·036$	$+0·018$	$+0·70$
16	15_3	7	$-0·002$	$+0·037$	$+0·018$	$+0·70$
17	17_2	8	$-0·022$	$+0·109$	$+0·012$	$+0·47$
18	17_2	7	$+0·045$	$0·000$	$+0·030$	$+1·17$
Mean			$+0·082$	$+0·022$		$+0·86$

Figure 2.3. The orderly presentation of scientific data: tables summarizing results from the Brazilian astrographic telescope in Eddington's eclipse expedition of 1919.

and ultimately dropped from the textbooks altogether, leaving the more Einsteinian numbers supplied by the other Brazilian telescope and the Príncipe telescope to decide the issue conclusively against Newton and in favor of Einstein's theory of relativity.

In the methodist's dream of science, the bodies of data from the three telescopes, the three measurements of gravity's power to bend light, would be assessed by a procedure that evaluated the evidential weight of each as carefully and as coldly as Eddington had in the first place calculated the numbers. The method would act, in effect, like a high-minded tribunal, objective and authoritative, sorting truth from falsehood without playing favorites or allowing personal or moral or self-aggrandizing considerations to enter into its deliberations.

If the Eddington story is any guide, this is pure mythology. There was no tribunal, no method, to sort the good photographic plates from the bad. The matter was settled the old-fashioned way, by a mix of partisan argument, political maneuvers, and propaganda.

SCIENTIFIC TRIBUNALS MAY be uncommon, but they have been assembled on an ad hoc basis from time to time, and one in particular offers some signal lessons about science.

Louis Pasteur is perhaps the most renowned of all French scientists— and surely the most revered by the French themselves. In his lifetime, from 1822 to 1895, he pioneered vaccination against anthrax and rabies, helped to discover the nature of fermentation, developed a sterilization technique ("pasteurization") to prevent milk and wine from spoiling, laid the foundations for the germ theory of disease, and uncovered the first evidence for the remarkable fact that the chemistry of life is overwhelmingly composed of "right-handed" molecules.

A few years ago, while visiting the École normale supérieure in Paris, I was given the privilege of using Pasteur's old office for a few weeks. (Pasteur served as the scientific director of the ENS from 1858 to 1867; at one point he banned smoking at the school, whereupon almost every student resigned.) Sitting at the antique desk, hoping for the greatness that lingered in that room to diffuse through my nerves and into my fingertips as I typed, I would occasionally be interrupted by visitors knocking at the door, eager to breathe in the august atmosphere of nineteenth-century experimentation and discovery. For the French, Pasteur is scientific thinking made flesh.

One of Pasteur's great victories was the refutation of the doctrine of spontaneous generation. Boil hay in water and decant the resulting fluid into an airtight container. Nothing happens. But let in a little air, and mold begins to grow. Where does it come from? Some nineteenth-century scientists held that the inanimate matter in the hay infusion reacted with the air to form, spontaneously, life where there was none before. Pasteur held, to the contrary, that with the air from the outside came dust containing invisible "germs" or "spores" of mold, which took root in the infusion. To nurture life, the solution had to be seeded with life.

It was clear enough in principle how to decide between these two opposing views. Introduce air that is free of "dust" or "spores" to the mixture. If life develops, spontaneous generation is real.

In practice, the problem is knowing that you have successfully found or created sterile air, given that the stuff we breathe looks much the same with or without spores. Many ingenious solutions were proposed. Air was heated or passed through acid to kill the spores. Experiments were conducted in long-neglected archives, where all dust was supposed to have long ago settled. Air was stored in a container coated with grease to trap the dust or passed through a long and sinuous tube that was supposed to perform the same function (Figure 2.4).

Figure 2.4. Swan neck flask.

The most scenic route to dust-free air was up a mountain trail. In 1860, Pasteur took 20 carefully prepared infusions to the Mer de Glace, a glacier on the Mont Blanc massif in the French Alps, where he exposed them to a chill, pure alpine wind more than 6,000 feet above sea level. Back in Paris, only one developed a moldy growth. Air alone, it seemed, could not bring organisms into existence. But Pasteur had competition. His great rival Felix Pouchet retaliated by performing the same experiment high in the Pyrenees, and all 8 of his infusions, on return to sea level, sprang to life.

Pasteur and Pouchet had sparred the previous year over their contrary views about spontaneous generation. A committee of the French Académie des sciences was convened to issue a prize for the best experimental investigation of the question—a competition whose outcome was understood by both parties to constitute a definitive verdict on the possibility that slime and mold might be created as a matter of course from inorganic ingredients. When the committee assembled, Pouchet discovered that it was packed with allies of Pasteur. He withdrew rather than face such a suspect tribunal. Now, after the success of his Pyrenees experiments, he and Pasteur negotiated a rematch. Once more Pouchet turned up only to find that the judging committee was composed entirely of opponents of his theory. He suggested a change to the rules, which Pasteur persuaded his friends to resist. Again Pouchet withdrew. That was the end of spontaneous generation.

Some writers have accused Pasteur of ensuring that both tribunals were stacked; they point to his reputation (which has been somewhat tarnished by the recent release of his laboratory notebooks) as a combative and unfair disputant in scientific argument. I understand the episode rather as a kind of real-life parable, illustrating the fact that in the scientific process, the weighing of the evidence—the tribunal's task—is seldom objective, seldom particularly methodical, always open to personal and political influence, and ever issuing decisions that are guided as much by expedience as by logic.

My story so far has relied largely on case studies—on anecdote, if you will—but the moral is brought home by several rather unsettling examinations of industry-sponsored research.

Commercial interests sometimes fund independent scientific investigations in the hope of turning up facts conducive to their profits. It emerges that of two groups of scientists working on a question, one financed by industry and one not, the industry-supported group is considerably more likely to produce commercially favorable findings, even when that group consists of university scientists not affiliated with the industry in any other way.

Researchers funded by Coca-Cola, PepsiCo, and other soda manufacturers have been five times more likely than others to find that there is no connection between drinking sugar-sweetened soda and obesity. Those funded by cigarette companies have been seven times more likely than others to find that secondhand smoke has no deleterious effect on health. And whereas non-industry-funded investigators of the efficacy of new drugs may find that the drugs do what they are supposed to do in about 80 percent of studies, investigators funded by the drugs' creators find a positive result almost 100 percent of the time. It looks as though something is guiding the science above and beyond cold, hard facts, something that closely shadows cold, hard cash.

How is it that science, for all its protocols and procedures and statistical handbooks, remains so malleable, so subject in its deliverances to personal, social, financial interests? Do scientists knowingly and deliberately subvert or ignore the scientific method, saluting it in public but then in private doing whatever best suits their ends? Or is the scientific method itself a unicorn, a name for something that isn't really there—leaving scientists to muddle through as best they can using the same rules of thumb that humans have relied on for millennia, subject to all the usual prejudices? Neither explanation, I think, is quite right.

But the lesson is clear: the outcome of the scientific process is powerfully influenced by the aims and interests of its practitioners, from

Eddington's desire for a European peace to the more utilitarian concerns of researchers funded by Big Tobacco or Big Soda. That is one sense in which science is, in spite of the methodists' hopes, decidedly subjective.

In the cases I have described, the "subject" in "subjective"—the researcher—imposes, wittingly or otherwise, their own goals on the course of science. Alongside this steering of science toward personal ends, there is another quite separate sense in which science is run through with subjectivity: scientists impose on science not only their goals but also their theoretical and explanatory tastes. Let me tell another story. It begins with a brand-new book.

BEFORE THERE WAS the internet, nothing said "information" so loudly and lavishly as an intricate map—except perhaps a volume of maps. In 1911, a young German scientist and explorer named Alfred Wegener looked into such a volume, a new edition of Andree's venerable *Allgemeine Handatlas* that made available for the first time the measurements of the ocean's depth conducted by the British *Challenger* expedition. What Wegener saw in the atlas astounded and provoked him.

Many previous geologists, cartographers, and map lovers had noticed the curious similarity between the coastlines of South America and Africa, which suggested that they had been cut from the same colossal slab of rock and dropped on opposite sides of the Atlantic. But nineteenth-century geographers knew that sea levels had risen and sunk considerably over the ages. When the sea level changes, the shape of the coast changes; coastlines, then, are ephemeral, and so the fact that there right now happens to be a suggestive match between the American and African seaboards tells you nothing much about the ultimate origins of those continents.

Thanks to the *Challenger* data, the new German atlas was able to show the outlines not only of landforms but also of continental shelves, those underwater extensions of continents in virtue of which offshore

waters are relatively shallow—until they plunge suddenly to truly oceanic depths. The forms of the continents plus their shelves are fixed in a
way that coastlines are not: they do not change as the sea rises and falls.
What Wegener saw in the pages of the atlas was an almost perfect match
between the eastern continental shelf of South America and the western
shelf of Africa. Such a match, he thought, could be no coincidence.

Inspired, he wrote what was to become one of the most controversial
books of the new century. *The Origin of Continents and Oceans*, which was

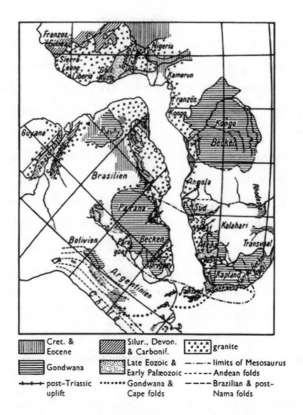

Figure 2.5. A map from a later edition of Wegener's *Origin of Continents and Oceans*,
based on the work of the South African geologist Alexander du Toit, showing the
geological and fossil (*Mesosaurus*) continuity in the South American and African
landmasses.

published in 1915, the same year as Einstein's relativistic theory of gravity, drew on geological and paleontological evidence to argue that Africa and South America must once have been nestled together in an intercontinental embrace. Not only were their shelves a perfect fit; a number of rock formations and fossil remains in one continent left off at the margins of the Atlantic only to pick up again on the other side, in just the place you would expect if the two continents had originally been a single landmass (Figure 2.5). How, then, did they come to be separated by one of the world's great oceans? In some way, suggested Wegener, the continents must have found a way to move over the surface of the earth. Thus was born the theory of continental drift.

The notion of millions of square miles of mountain, forest, river valley, desert, and steppe sailing blithely through the ocean on a voyage to parts unknown is even today difficult to take on board. Skeptics could not accept Wegener's theory without a persuasive hypothesis about the mechanism of drift, a hypothesis that was both plausible and for which there existed some direct evidence. Wegener offered only unsubstantiated speculation about the mechanism—he had no more idea than the skeptics how drift occurred—but he thought that the evidence that it had happened, one way or another, was overwhelming. The fate of the theory teetered: in some places drift was looked on with favor, in others with doubt verging on incredulity. In 1943 it was attacked, however, by the eminent evolutionary biologist George Gaylord Simpson, and it sank under the weight of his reputation. Wegener was by then unable to defend himself. He had perished in 1930 in the most adventurous of circumstances, on a desperate mission to resupply a scientific outpost in Greenland in the face of the oncoming winter. Though he had enjoyed a successful scientific career, the scorn directed at his ideas about drift left him, at his death, a disappointed man.

Why did Wegener fail to persuade the world? He was quite right to think that his accumulated evidence reflected the diverse effects of continental drift. But his critics, including Simpson, were right to think that the lack of a well-evidenced mechanism for such a titanic process was a major consideration against the theory. Who was more right? As so often in science, both the pros and the cons were persuasive.

If the continents move, there must be a physical process by which they accomplish this feat. Wegener's contemporaries knew of no such mechanism. Those who wrote against drift thought wrongly that they understood enough about the earth's geology to rule out any plausible physical story about drift. Those who wrote in its favor were prepared to take the chance that a mechanism existed, though they could only make guesses about its nature. The opponents of drift were overconfident; the proponents were extraordinarily bold, perhaps reckless.

The most reasonable response to Wegener's ideas, you might think, would be simply to sit out the dispute. That would be a mistake. Scientific judgments must sometimes be made using decidedly incomplete evidence—inaction may be disastrous, as now seems to have been true in the case of global climate change. More important still, as both Kuhn and Popper urged, in order to test a theory properly, to dig out all the most telling evidence both for and against it, you need partisans. Only those who are committed to proving the theory true—Kuhn's paradigm-bound scientists constitutionally incapable of doubt—or to proving it false—Popper's agents of generalized theoretical destruction—will have the motivation to perform years or even decades of necessary experimental work. If every scientist had reacted to Wegener by deciding to "wait and see," we might still be waiting.

So scientists must make decisions as to how the evidence bears on theory when there are compelling arguments on both sides. It comes down, often enough, to personal taste—or perhaps to professional circumstances. An establishment man like Simpson—a professor at Columbia and Harvard and a winner of several prestigious awards around the time he set out to refute the hypothesis of continental drift—may feel more at home with the apparently safe or orthodox view, while an outsider and adventurer like Wegener—who was among other things a record-breaking balloonist—may be more ready to take the risk that what's vital to his theory lies just over the scientific horizon. Sometimes the decision is taken on shallow practical grounds: scientists in the United States, where Simpson was more powerful, tended to be far more skeptical about drift than those in Britain and Europe.

In making their contrary calls of judgment, Simpson and Wegener did not ignore the available evidence, but they were heavily influenced also by their temperament, their social position in the institution of science, and no doubt countless other elements of the psyche that differ from one mind to the next. That's how human thinking works. So we get splendid variety; so we get bitter dispute.

ACCORDING TO THE THINKERS I call methodists, what makes science special is a standardized rule or procedure for conducting empirical inquiry. If what you have seen of science in this chapter so far is representative, nothing of the kind is true. It is not method that conquers human frailty; it is human frailty that conquers method. Just when objectivity matters most, scientists—great scientists, perhaps, above all—are apt to draw on their deepest rhetorical and political resources to skew the course of inquiry to favor their own ends. There is no higher authority to curb the chaos. Science has no single impartial judge to hand down decrees to which all researchers are obliged to subscribe, but rather numerous contending Pasteurian tribunals, each packed with the advocates of a particular set of interests or a particular way of seeing the world.

I began *The Knowledge Machine* by presenting the ideas of two great methodist thinkers not only because methodism is simple, appealing, and historically important, but also because I, too, am a kind of methodist: I want to appeal to a shared scientific code to account for science's supreme ability to find theories of great predictive and explanatory power. In that case, I seem to have run into something of a problem: a code is not much use if it is ignored just when it matters most. Chapter 2 is not yet over and methodism is already battered, down on its knees, begging for mercy.

To the rescue comes an idea entertained by many concerned scientists, advanced by Karl Popper, and articulated here by the writer and surgeon Atul Gawande:

> Individual scientists . . . can be famously bull-headed, overly enamored of pet theories, dismissive of new evidence, and heedless of their fallibility. . . . But as a community endeavor, [science] is beautifully self-correcting.

That "science is self-correcting" is often heard in the wake of revelations of fraud or methodological recklessness. Those who mount this defense of scientific objectivity, such as Gawande, ruefully concede that the bad or at least careless behavior of scientists such as Eddington or Pasteur is a serious impediment to scientific progress: it may slow, divert, even temporarily reverse the growth of knowledge. But scientific inquiry is both competitive and cooperative enough that its practitioners check up on one another, hoping either to demonstrate the reliability of research they plan to build on or to debunk work that conflicts with their own results. Even if scientists lack the proper Popperian critical perspective on their own work, they have many reasons to apply that exacting and skeptical attitude to the work of others.

Further, in the opinion of most "self-correctors," the transgressions documented in this chapter are real but unrepresentative. They are the most extreme cases, selected for anecdotal rather than statistical effect. Were you to raid some science lab at random, you would find goings on far less salacious: not saintly by any means, but for the most part a creditable attempt to more or less follow the objective rules of inference laid down by the logic of the scientific method.

As a matter of fact, maybe a visit like that is not such a bad idea.

IN 1975, a young French anthropologist named Bruno Latour went to live among the natives of an unusual southern Californian subculture. His subjects were researchers working in the laboratory of the endocrinologist Roger Guillemin—the scientist who was to share the Nobel Prize in Medicine two years later for determining the structure of the brain hormone TRH.

At the time, Latour's "knowledge of science was non-existent," he later wrote—adding that he was "completely unaware of the social studies of science"—but for these very reasons he was "in the classic position of the ethnographer sent to a completely foreign environment," like Margaret Mead in the Samoan archipelago or Napoleon Chagnon among the Yanomami of the Amazon rainforest.

Latour spent two years inside Guillemin's knowledge machine at the Salk Institute in San Diego. He found a great wheel, a self-replenishing cycle of being. Into the Guillemin lab came vast quantities of chemicals, animals, and energy, which fueled a complex physical and social process that produced scientific reports by way of "inscription devices" whose function was to "transform pieces of matter into written documents" (Figure 2.6). These documents, the articles published in scientific journals, were in turn transformed into "credit," or scientific reputation, which is valuable largely not as an end in itself but because it provides the means to win funding that will buy more chemicals, animals, and energy, along with new inscription devices and the services of scientists and technicians to turn them into more papers and thus still more credit. A science lab, in the telling of Latour and his collaborator Steve

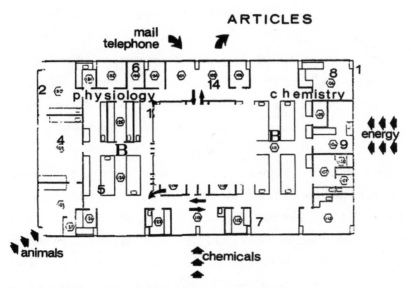

Figure 2.6. Guillemin's knowledge machine, as annotated by Latour.

Woolgar in their book *Laboratory Life*, is not so much an appliance as an organism: its primary concern is survival and reproduction.

What about the rules of thought that, according to Gawande and Popper, govern most ordinary scientific inquiry—the code of intellectual conduct against which scientists' reasoning is periodically compared, allowing science to "self-correct" and so to stay on a more or less objective path?

Latour did see much that was objective in the Guillemin lab. Exacting procedures for pulverizing brain tissue and extracting substances such as TRH were known to all the technicians and were carefully followed. The same is true of the preparation of the data that was used to test hypotheses about TRH and other molecules' structure. In one such technique, a synthetic substance with the hypothesized structure was created and compared with TRH, to see whether there was a match. The comparison involved images created using the two substances, called "chromatograms." If two (correctly prepared) substances have the same chromatograms, they have the same components; identical chromatograms, then, show that the hypothesized structure is correct. A chromatogram is prepared using a machine that can be ordered from a scientific supply company, and that machine comes with a manual. In the manual are instructions for using the machine correctly; if a scientist fails to follow the manual, they have erred objectively and conspicuously, a mistake that can be corrected, just as Gawande and Popper maintain.

But although the preparation of evidence may have answered to objective rules, Latour found that the interpretation of evidence did not. There will always be small differences between two chromatograms of the same substance, in the same way that there are small differences in two photographs of the same person. Scientists must decide, then, when those differences are small enough not to matter. If there was a rule to decide these questions, Latour's subjects did not follow it. They resorted, rather, to "local tacit negotiations, constantly changing evaluations, and unconscious or institutionalized gestures." The same was true of all other important questions of interpretation: in deciding how the evidence bore on the hypotheses that it was supposed to test, there was no appeal to shared rules, to objective criteria. Instead there were arguments, gut reactions, bargains, local cultures. Summing up, Latour and Woolgar wrote, "We were unable to identify explicit appeal to the norms of science."

In Guillemin's lab, then, there was much objective weighing of brains and their juices, but little objective weighing of evidence. When it came to determining what the carefully collated data said about theories of TRH's structure and the like, not even the methodological yeomen—the postdoctoral fellows, the junior scientists, the quiet, decent majority flying in science's economy class—followed rules. Heedless of official restrictions, they went on stuffing the overhead bins of scientific inference with their moral, psychological, political, and cultural baggage.

It is the way things are throughout the scientific world. The hematologist James Zimring reports the dismay of novice scientists upon encountering for the first time the reality of the lab:

> The work they were doing and that of their fellow researchers seemed "messed up." It was chaotic, did not progress logically. . . . Often rationalizations were made up after the fact to account for progress.

Science can hardly correct itself if no one is paying the least attention to standards for correctness.

Where is science's method, then? It is nowhere at all, say the followers and fellow travelers of Latour. There is no structure or system that makes science a more objective, a more valid, a more truth-directed way of knowing the world than any other. The Great Method Debate is an argument over a fiction. That is the thesis, endorsed by many contemporary sociologists and historians of science, that I call *radical subjectivism*. It is the antithesis of methodism.

According to the radical subjectivists, then, the world of scientific inquiry is, for all its specialized apparatus and ideology, essentially a microcosm of the multiplicity of human society, its tens of thousands of participants each having their own idea of what is worth doing and how it might be done, traveling more often at cross-purposes than together, sometimes not talking at all, sometimes arguing with each other, sometimes subtly undermining each other, sometimes seeing only what they want to see, sometimes seeing only what they've been told they'll see, sometimes only seeing their status relative to their rivals in an endeavor whose content, the stuff of scientific theories, may be treated more as a means to self-promotion than as an end. Science, in the radical subjectivist view, is just another venue for the Machiavellian masterpiece theater that is the human condition.

What about the force of the empirical facts? "The natural world has a small or non-existent role in the construction of scientific knowledge," writes the sociologist of science Harry Collins; similarly, the sociologist Stanley Aronowitz says, "Science legitimates itself by linking its discoveries with power, a connection which *determines* (not merely influences) what counts as reliable knowledge." The facts are, in short, pawns in a game in which the "strongest team decrees what counts as truth."

Popper's focus on falsification, Kuhn's paradigm-by-paradigm approach, any other system or method that asks scientists to put aside their mortal ways: hopeless, each and every one, according to the radical subjectivists.

Scientists are all too human. They do not follow a universally valid script. They do not follow the locally valid script of a Kuhnian paradigm. They are irrepressibly contentious, contextual, and quirky.

The radical subjectivists are, I think, right about the subjectivity of science. But they cannot be right in their further claim, that there is nothing whatsoever to distinguish science from ordinary thought or philosophical contemplation. That would explain everything about the messy human business of scientific inquiry except what matters most: the great wave of progress that followed the Scientific Revolution. Medical progress, technological progress, and progress in understanding how it all hangs together, how everything works. Immense, undeniable, life-changing progress.

Some radical subjectivists hint that the supposed contribution of scientific understanding to the advance of human happiness is mere propaganda put about by Big Science, arguing that technological advances issue more from trial and error than from deep knowledge of the underlying structure of things. But most acknowledge science's theoretical as well as its practical success. "Science remains . . . certainly the most reliable body of natural knowledge we have got," writes the sociologist Steven Shapin. Such an abrupt conclusion to the historical and sociological litany of scientific fallacy, frailty, and confabulation adds up to something not unlike this retelling of the famous fairy tale: *Cinderella was a poor and bedraggled child, beaten and abused by her stepmother and stepsisters, treated little better than a slave. And so she married the handsome prince and lived happily ever after.* A part of the story must be missing—the best part.

Among the detritus left behind after the ball—among the ulterior motives, self-serving explanations, power plays, crushed flowers, shattered beakers, broken promises—you might still hope to spot the gleam of a glass slipper, the clue that will join together the two halves of the tale. The slipper is a scientific method, an objective rule that lays down a standard of scientific conduct followed even by the Eddingtons and the Pasteurs, answering the big philosophical question about the source of science's knowledge-finding power.

If it exists, the slipper must be a subtle and exquisite thing. It must fit with everything said previously about the malleability of scientists' words and actions while providing an alternative to the radical subjectivist interpretation. It must show that scientific inquiry is not human business as usual, but rather that there is something unique and objective about the conduct of science that explains its success. It must permit but at the same time overcome our humanity.

Is that an impossible demand? I will show you that it is even more difficult to satisfy than you might have supposed. Not only human nature but also the very laws of logic contend against the possibility of objectivity in scientific thought.

CHAPTER 3

The Essential Subjectivity of Science

ℰℵ

The logic of scientific reasoning is by its very nature subjective.

I T TOOK THE JURY little more than an hour to find Todd Willingham guilty of setting the fire that killed his three daughters. The evidence was overwhelming. An arson specialist testified that many features of the fire could have been caused only by the laying and lighting of a trail of an accelerant, such as gasoline or charcoal lighter fluid, through the house ending at the front porch. A suspicious patch on the porch did indeed test positive for lighter fluid. Witnesses—neighbors and a fire department chaplain—testified that Willingham seemed strangely unperturbed as he watched the fire burn. And a prisoner jailed along with Willingham after his arrest testified that the accused had confessed to the crime, saying that he took "some kind of lighter fluid, squirting [it] around the walls and the floor, and set a fire." In August 1992, eight months after the con-flagration, a Texas judge sentenced Willingham to death.

A decade later, as Willingham languished on death row, things were beginning to look rather less certain. A sympathetic prison visitor, Eliz-abeth Gilbert, had found certain discrepancies in the onlookers' testi-mony: their interpretation of Willingham's behavior at the time of the fire had changed significantly for the worse after they learned he was being charged with murder, a cognitive effect well known to researchers. Around the same time, the prisoner who claimed to be a recipient of Willingham's spontaneous confession filed a "Motion to Recant Testimony." And perhaps most significant of all, a greatly increased understanding of the dynamics of house fires revealed that what were considered telltale signs of arson at the time of the trial could easily be caused by entirely accidental blazes. The lighter fluid on the porch? Most likely left by a charcoal grill destroyed in the inferno.

No one will ever know whether Todd Willingham murdered his daughters. He was executed in 2004. In 1991, those who contributed to the case against him—the fire marshals, the neighbors, the chaplain, the prosecutor—took themselves to be approaching the case responsibly, sincerely, and without prejudgment. But many now believe that the state of Texas killed an innocent man.

The criminal justice system strives to uncover the truth. Even when it operates as it should, however, its interpretation of the evidence may depend on whether a witness is reliable or a theory—such as the arson investigators' assumptions about the effect of accelerants—is correct. At the moment when it matters most, there may be no objective basis for answering such questions. Information is limited, yet a determination must be made. The deliberators have no choice but to fall back on what seems most plausible to them. Much later, it may become clear that a witness was untrustworthy or that a theory was flawed. Lacking a crystal ball, the jury must do its best with what it has at the time.

It is the same in science. Sometimes it is a measurement instrument— a witness, if you will—on which the issue hinges. Sometimes, it is a theory. Scientists seeking to make sense of the evidence cannot be neutral. They must take a stand on whether the instrument is relaying the truth, on whether the theoretical assumptions hold. Having nothing further to guide them, they must go with whatever seems right. They must resort to educated guesswork, and that makes scientific reasoning irreducibly, unavoidably, essentially subjective.

CAST YOUR MIND BACK to 1919, the year of Eddington's eclipse. The rationale for the expedition to observe the eclipse was straightforward. If Einstein was right, then the light of stars close to the sun would be bent by twice as much as if Newton was right. Measure the degree of bending, then, and you will see which of the two theories is correct.

In Brazil, two members of Eddington's team focused their astrographic telescope on the eclipse and took 18 photographs. The results of those photographs are summarized in Figure 2.3, where (as you will remember) the observations are condensed into the single number in the bottom right-hand corner, showing the overall degree of bending: an almost perfectly Newtonian 0.86 arc seconds. Eddington protected his Einsteinian agenda, however, by dismissing the significance of the astrographic telescope's photographs.

In so doing, it might seem, he violated the methodological commandment that Popper made famous, the precept that a theory making false predictions must be spurned. But that is not quite correct. As we will see, Eddington broke no objective rules in his belittling of the Brazilian data. Not even the most unscrupulous scientist could have done so, because the sociologists Bruno Latour and Steve Woolgar are right: no such rules exist. Further, this is not merely a matter of sociological fact but of philosophical principle. The most concerted attempts to frame such rules, such as Popper's principle of falsification, for systematic reasons fall through, ultimately putting no objective constraint on scientists' interpretation of evidence.

To see why, take a closer look at Eddington's rationale for ignoring the data from the astrographic telescope. He argued—controversially, but not arbitrarily—that something had gone wrong with the telescope, resulting in its systematically underreporting the gravitational bending. Although he had no direct evidence for this claim, there are a number of ways such "undermeasurement" might have occurred.

You cannot simply look through a telescope and see light bending. Rather, what you must do is photograph the apparent positions of the stars next to the sun's disk at the moment of the eclipse and compare them to the same group of stars when they are nowhere near the sun. The degree of bending is revealed by the difference in positions on the photographic plates. This difference is microscopic: the measured gravitational bending of 0.86 arc seconds is equal to 0.0002 degrees, which corresponds to a shift in position of only 1/60 of a millimeter—0.0007 inches—on the Brazilian astrographic's plates. Anything that has the slightest impact on the measurements will result in a significant error in the calculation of the bending.

There were many such potential spoilers, because the setup of the apparatus in Brazil was rather complex. In Figure 3.1 you see the astrographic telescope at home in the Greenwich Observatory in England. It is attached to a heavy, precisely engineered mount that allows it to be trained on any point in the sky. Eddington left that mount behind. In Brazil the telescopes were laid flat, pointing at the horizon (Figure 3.2). For each telescope, an external mirror reflected light from the target in the sky down the prone telescope's barrel.

Eddington and his team seized

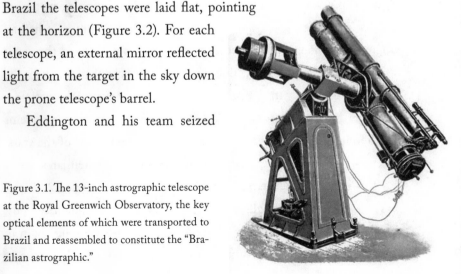

Figure 3.1. The 13-inch astrographic telescope at the Royal Greenwich Observatory, the key optical elements of which were transported to Brazil and reassembled to constitute the "Brazilian astrographic."

Figure 3.2. The setup of the Eddington expedition's telescopes in Brazil. The astrographic is on the left, the 4-inch on the right. The external mirrors are mounted on the block in the foreground.

on certain disadvantages of that arrangement to explain how the astrographic telescope's measurements might have gone wrong. The heat of the tropical sun beating down on the telescope's mirror before the onset of the eclipse, they conjectured, might have caused irregular expansion that distorted the photographic images. The mirror in any case had an astigmatism, though the scientists had found a way to avoid the worst consequences of this imperfection. Finally, the mechanism that kept the mirror pointing toward the sun, compensating for the earth's rotation, was operating irregularly. It would not be so difficult for these problems to introduce errors of 0.0007 inches or so in the positions of the stars, errors that would deliver Newtonian numbers from Einsteinian skies.

The "empirical fact" reported at the bottom right of Figure 2.3—the gravitational bending angle—is not, then, an observed quantity but a calculated quantity, a number whose value depends on a long chain of assumptions, some of which might easily be false. The same is true of the bending angle measured using the 4-inch telescope that lay along-side the astrographic instrument, taking in the same patterns of light but announcing a contrary verdict. Indeed, in retrospect, we know that something was systematically off kilter in the 4-inch telescope as well, since it indicated a bending angle rather larger than Einstein's theory allowed. Eddington had to make a choice. Discount the astrographic data? Overlook the 4-inch discrepancy? Declare the experiment to be inconclusive? He did not have enough information to single out an obviously correct answer. So he followed his instincts.

Eddington's situation was not at all unusual. In the interpretation of data, scientists often have great room for maneuver and all too seldom have unambiguous guidance as to which maneuvers are objectively right and wrong.

The room for maneuver exists because, as the eclipse experiment shows, theories in themselves do not make predictions about what will be observed. To say anything at all about the experimental outcome—about, say, the position of spots on a photographic plate—theories must be supported and helped along by other posits, other presumptions about the proper functioning of the experimental apparatus, the suitability of the background conditions, and more.

In other words, a theory, like a medieval knight, never fights alone, but rather rides into empirical combat with a retinue of assumptions. It is this formation as a whole—what you might call the *theoretical cohort*—that makes predictions about and gives explanations of the outcomes of experiments, measurements, and other observations. The theory gets all the attention. But it cannot engage the enemy without its coterie of men-at-arms.

Consequently, when something goes wrong, a theory can be saved from refutation by blaming the assumptions—as Eddington did when he used his considerable logical, social, and political skills to have the Brazilian astrographic measurement, the patently un-Einsteinian 0.86

arc seconds, dismissed on the grounds that something had gone wrong with the apparatus. Faced with a faulty prediction, a scientist must decide when to sacrifice an assumption to save a theory and when to accept that the theory itself has failed.

Karl Popper took this problem very seriously. He had no choice, as it seemed to undermine his central idea, that science progresses by eliminating theories that make false predictions. If a theory can be excused whenever something looks wrong by blaming an assumption—by postulating some error in the measurement apparatus, for example—then how can theories ever definitively be let go?

Popper allowed that blaming an assumption to save a theory is sometimes the right thing to do, but only under certain conditions. He required that the new assumptions made in the course of such a defense should themselves be falsifiable and that their proponents ought, in the critical Popperian spirit, to make every effort to test them. This recommendation scientists are often happy to follow. In late 2011, neutrinos created at the CERN research facility in Switzerland were clocked traveling faster than the speed of light—an athletic feat forbidden by Einstein's theory of relativity. Rather than discard relativity, the great majority of physicists supposed instead that something had gone wrong with the measurement apparatus. The matter did not rest there, however; having saved relativity from falsification, they followed Popper's advice and set to work testing the supposition on which the rescue depended: that "something had gone wrong." An exhaustive overhaul of the experimental machinery vindicated their conservatism. It turned out that a cable was loose.

Such care and attention, however, is not always feasible. Back in England, writing up his results months after the eclipse, there was no way that Eddington could double-check the effect of, for example, the mirror's expansion in the sun's heat that day. The same is true for many suspected experimental malfunctions: the supposed aberration is often temporary, and there is no way in retrospect to determine whether it happened or, if it did, to what degree. The facts of the matter are lost to history.

Popper suggested a different way to deal with these cases: do the experiment a second time, more carefully. Again, scientists are readily observed following this advice—but again, it is not always feasible. Solar eclipses are rare enough; what made Eddington's 1919 eclipse rarer still was the sun's position, at the time of totality, in the center of a field of relatively bright stars. As Eddington pointed out when touting the experiment, this happy alignment "would not occur again for many years." He might have wanted to go back for another round of stellar photography, but he could not—so he found other ways to press his case against the Brazilian astrographic and in favor of Einstein.

Eddington's course of action was unconstrained not because he disdained the rules of scientific thought but because the complexities and difficulties of empirical investigation—of making precise measurements of small or barely accessible quantities—meant that he had no rule capable of telling him how to interpret his photographic plates. Even the tenacious attempts of that great methodist Karl Popper to lay down principles for deciding, in the face of a faulty prediction, whether to blame the theory or merely a measurement instrument were of no help. Sometimes a scientist striving to interpret the significance of empirical data, like a jury member faced with questionable testimony, simply has to make a judgment call—personal, instinctive, subjective.

Eddington's and Pasteur's self-regarding maneuvers, Latour's ethnographical investigations—these were bad enough for methodism, showing that scientists both famous and obscure fail to follow an objective guide when assessing the impact of their evidence. Now the situation appears positively dire: in many cases, there are no such guides. Not even if science were flawless, populated by paragons of temperance, rationality, and selflessness, could it assess evidential weight objectively.

Or at least that is true when the significance of the evidence depends on the contested credibility of a measuring instrument. It is also true, just as in the courtroom, when what the evidence says depends on the plausibility of a controversial theoretical assumption—as we'll now see.

GEOLOGISTS BEGAN TO fathom in the early 1800s that the earth is extraordinarily ancient. Charles Darwin's theory of evolution by natural selection, proposed in 1859, needs as many millennia as it can get its hands on in order to provide time enough for all the diversity and complexity of life to emerge from a common ancestor—for flecks of floating protoplasm to sprout branches and leaves, heads and legs, and to take over the planet's surface. Darwin therefore seized on the new geology to argue that the earth must be at the very least hundreds of millions of years old.

This staggering idea ran full tilt into a formidable obstacle. The name of the obstacle was William Thomson, one of the most famous and influential physicists of the time. Thomson was a prodigy: born in Belfast in 1824, he published three scientific papers while still in school and at the age of 22 was appointed a professor at the University of Glasgow, where he remained his entire life. He made important discoveries in the new sciences of energy and heat and pioneered the notion of the heat death of the universe—the inevitable dispersion of energy that would result in the world's becoming a quiet, dark, homogeneous, and lifeless place in which everything was at the same temperature and nothing more could happen. Turning to engineering and commerce, he joined the effort to lay an undersea telegraph cable between Britain and the United States; after years of accidents and false starts, the connection was made in 1866 and Thomson was knighted for his contributions. He then headed the opposition to Irish home rule in the Liberal Party, a service for which he was ennobled in 1892, becoming Lord Kelvin—the name by which he is usually identified today.

Kelvin was throughout his life conventionally religious. He was a latitudinarian, meaning that he considered distinctions between denominations—in particular, between Anglicanism and Presbyterianism—to be unimportant; indeed, as a source of revelation, he preferred nature to the pulpit. Look around and you will see at work "a creating and directing power," he wrote, and so "if you think strongly enough you will be forced by science to the belief in God, which is the foundation of all religion." But the science of Darwinism, as he understood it, threatened to broadcast exactly the contrary message.

Kelvin had long hoped to calculate the age of the earth using the

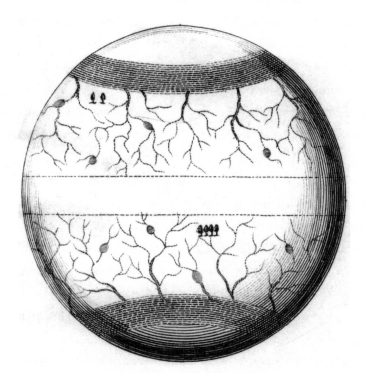

Figure 3.3. The earth at the dawn of creation, according to Thomas Burnet's *Sacred Theory of the Earth*, published in 1681. Rivers flow from the poles to the equator. A speculative location for the Garden of Eden is marked by a line of four trees in the southern hemisphere.

physics of heat. His idea was straightforward. The colder a cup of coffee, the longer it must have been sitting out on the counter since it was poured. Likewise, the colder the earth's crust, the longer the planet must have been cooling since its formation. The earth's age could be estimated, then, if both its original temperature and the current temperature of its outer layer were known. Kelvin presumed the original temperature to be that of molten rock, but for the current temperature he had to wait some years, until the Scottish physicist and glaciologist J. D. Forbes made a series of measurements of the temperature of the subsurface rock around Edinburgh.

With these numbers in hand, Kelvin published his calculations in 1863 using a well established theory of cooling to show—by the time he gave his final estimate in 1897—that the earth could be no more than 20 to 40 million years old and "probably much nearer 20 than 40." Its crust was too warm for the planet to have been cooling any longer. Further, other late nineteenth-century estimates of the earth's age based on the probable age of the sun, some made by Kelvin himself, also came in as low as 20 million years. It seemed that physics would give evolution no time to create life's variety. Darwin stood refuted.

To the rescue came Darwin's most tenacious defender, his "bulldog," the anatomist Thomas Huxley. In 1860, Huxley had famously seen off Darwin's critic Bishop Samuel "Soapy Sam" Wilberforce, proclaiming in the face of Wilberforce's trite mockery that he would rather be descended from apes than from a man who scoffed at serious debate. Huxley then fought a running battle with the paleontologist Richard Owen, who argued that the resemblances between the brains of apes and humans were merely superficial and therefore no evidence for their descent from a common ancestor. Now the bulldog's tactical skills were called upon once again.

Huxley did not, in truth, know a great deal about the physics of heat. But he knew how to win an argument, and so armed, he went to work on the plausibility of Kelvin's assumptions. Kelvin's mathematics was impeccable, Huxley acknowledged, but accurate calculation was not enough:

> What you get out depends on what you put in; and as the grandest mill in the world will not extract wheat-flour from peascods [pea pods], so pages of formulae will not get a definite result out of loose data.

Further, Huxley opined, Kelvin was a mere "passer-by" who did not understand the deep foundations of the geology and biology on which Darwin's theory was built. But geology is a branch of physics, Kelvin replied, and so as a physicist he was, far from being a passer-by, an expert whose opinion about its foundations ought to be taken very seriously indeed.

Kelvin's response was disappointingly polite, thought his friend and colleague Peter Guthrie Tait, who charged into the debate implying that natural historians such as Darwin and Huxley were "beetle-hunters" and "crab-catchers" incapable of recognizing the power of mathematical thought. He concluded with an estimate of the earth's age that undercut even Kelvin's: "Natural Philosophy [that is, physics] already points to a period of ten or fifteen millions of years as all that can be allowed for the purposes of the geologist and paleontologist; and . . . it is not unlikely that with better experimental data, this period may be still farther reduced."

The earth is actually over 4.5 billion years old, and it has harbored life for at least 3.5 billion of those years. How did Kelvin get it so badly wrong? Like Todd Willingham's jurors, who were presented with an inadequate theory of the way in which house fires develop and burn, he was relying on assumptions that were mistaken in several respects. First, though he had no way of knowing it, the heat of the rock making up the continents is considerably increased by the decay of radioactive elements. Second, heat is transported from the earth's core not by conduction through solid rock, as Kelvin had supposed, but by convection, in which rock in the earth's mantle flows from the core to the crust carrying heat with it, warming the crust far more efficiently than conduction could. As a consequence, an old earth can have, as ours does, a surprisingly warm crust.

In effect, Kelvin sipped the top layer of a cup of coffee and, finding it to be piping hot, concluded that it had been freshly poured. In fact it had been there for ages, but sitting on a warmer and continually stirred. Huxley was right. Kelvin had poured chaff into his finely calibrated mathematical mill and produced indigestible grit.

Were Kelvin, Tait, and other advocates of a "youngish earth" being unscientific? Not at all; there were good reasons and great uncertainty on both sides. The physicists saw their assumptions about the geological structure of the earth as natural and reasonable compared with the largely speculative theory of evolution; if something had to go, it was biological guesswork, not careful physical extrapolation. The biologists saw their grand explanation of life's intricacy, though hardly proven, as a breathtaking achievement based on extensive observation of nature across the globe, a breakthrough that could scarcely be discarded on the grounds of pure conjecture about the unseen goings-on thousands of miles beneath their feet. Such contrary attitudes are business as usual in science, an unexceptional manifestation of the subjectivity that swirls through all scientific reasoning, planning, and debate.

Karl Popper sought, by laying down rules of scientific method, to resolve disputes such as this, to decide whether observations of the earth's surface temperature falsified Darwin's theory or whether somewhere in Kelvin's thinking "something had gone wrong." But Popper's precepts were no more useful in the age-of-the-earth controversy than they were in the Eddington affair. When making theoretical assumptions, he said, be bold. Choose hypotheses that, by making strong claims, expose themselves forthrightly to falsification. Kelvin and Darwin certainly did that. But they had no way, in their lifetimes, of testing their claims.

Popper might perhaps have counseled both physicists and biologists to keep an open mind, to refrain from taking sides, until more was known and a definitive falsification of either biological theories or physical hypotheses was achieved. Such a prescription is hardly realistic: it is precisely the sort of stricture that scientists, being also humans, will consistently fail to follow. And in any case it is, as we saw in the case of Wegener and continental drift, bad advice. Science is driven onward by arguments between people who have made up their minds and want to convert or at least to confute their rivals. Opinion that runs hot-blooded ahead of established fact is the life force of scientific inquiry.

For these reasons, Popper is now thought by most philosophers of science to fall short of providing a rule for bringing evidence to bear on theories that is both fully objective and adequate to science's needs. What kind of rule might do better? There is philosophical consensus on this matter, too—and the answer is *none*. An objective rule for weighing scientific evidence is logically impossible.

The impossibility arises from the same fact that makes Popperian falsification often such a contentious matter: a scientific theory issues predictions only when it is combined with various assumptions to compose a theoretical cohort. The members of the cohort—what philosophers call "auxiliary assumptions"—are a diverse range of suppositions. Some are high-level theories themselves, such as Kelvin's assumption about the solid structure of the earth's interior. Some are assumptions about the functioning and calibration of measuring instruments, such as Eddington's assumptions, positive and negative, about his various telescopes. The auxiliary assumptions are like links in a chain leading from the theory to the evidence. The chain is only as strong as its weakest link; thus, to assess the strength of the chain—to assess the strength of a

Figure 3.4. To evaluate the overall strength of a scientific argument, you must evaluate the strength of each of the argument's pieces.

piece of evidence for or against a hypothesis—you must have an opinion about the strength of each of the links.

It is, in other words, impossible to judge the impact of a piece of evidence on a theory without having a view about the auxiliary assumptions. If you think that the Brazilian astrographic telescope was working perfectly, you will count its measurements of the bending angle of light as powerful evidence against Einstein's theory of relativity. If you find it quite plausible that something went systematically wrong in those measurements—if you suspect that this particular link in the evidential chain is faulty—then you won't regard the evidence as at all strong; you will, like Eddington, be inclined to overrule it on the basis of evidence obtained from other instruments in which you place greater trust.

Likewise, if you think that Kelvin's assumption about the solidity of the earth's insides is on firm ground, you will (provided you have some faith in his other assumptions) interpret nineteenth-century measurements of the temperature of the earth's crust as strong evidence for a youngish earth, and consequently as doing great damage to Darwin's theory of evolution. If, by contrast, you think of the solidity assumption as an exceptionally risky conjecture, supposing as it does that the rigid structure found in the top few miles of the earth's surface must continue down unchanged for another 4,000 miles to the planet's core, then you will consider the temperature measurements to be only piffling evidence against Darwin's ideas.

A rule that strives to lay down the law about the significance of scientific evidence, then, must also lay down the law about the likelihood of all relevant auxiliary assumptions, in the same way that a procedure for determining chain strength must estimate the strength of every link. The rule's judgments can be objectively valid only if its estimates of the auxiliary's likelihoods are objectively valid. An objective rule for weighing any and every piece of evidence is therefore possible only if there is an objective fact of the matter about the likelihood of each relevant auxiliary assumption, given the available evidence.

As we have seen, however, opinions about auxiliary assumptions can differ wildly—not because scientists ignore the rules of right reasoning, but because there are simply not enough known facts to nail down likelihoods for every auxiliary assumption in a theoretical cohort.

On the one hand, as the Eddington affair illustrates, assumptions about experimental conditions and the transient state of the measurement apparatus are often impossible in retrospect to check, and experiments and observations are often too difficult or too expensive to repeat—in the short term, at least.

On the other hand, to form an opinion about a theoretical auxiliary assumption, such as Kelvin's assumption that the earth is entirely solid, requires further evidence, and the significance of this evidence for the auxiliary assumption will itself depend on further auxiliary assumptions. Among these assumptions may appear the original hypothesis, forming an unbreakable circle.

When Louis Pasteur, for example, ventured to show in the 1860s that life could not form spontaneously from an inorganic mix of "hay soup" and air, he needed a supply of air that was sterile, that is, free of the "spores" that he hypothesized to be the source of all mold, slime, and other growth in the soup. As you may remember, he and other experimenters tried various procedures to obtain spore-free air: heating the air, storing it in a greasy container, sampling it from alpine peaks. That such air is indeed sterile is a classic auxiliary assumption, essential for the validity of the experiment. But how to ensure that it holds true? The only way that Pasteur knew of testing his auxiliary was to mix the air with the soup to see whether life developed; if it did, the air contained spores, and if not, not. But such a test of course assumes the very theory that Pasteur was trying to prove, that life could not emerge spontaneously from sterile air and soup. The experimenters on both sides of the spontaneous generation debate at the time had no way of independently verifying their most important auxiliary assumption.

There is no way to get started, in such a situation, without assigning some likelihoods from scratch—not arbitrarily, exactly, but without the constraints imposed by a preexisting scheme for interpreting evidence. Needless to say, different scientists will choose different starting places, heavily influenced by personal tastes or aspirations. From that point on their estimates of evidential weight are liable to head in disparate directions. That is the origin of the essential subjectivity of science.

Subjectivity need not mean anarchy. There are rules for interpreting evidence, but they are rules that allocate a role to subjectively formed estimates of a hypothesis's likelihood, or, as I will call them, *plausibility rankings*.

As an example, consider a precept that has already had a vigorous workout: a piece of evidence should count for less the more likely it is that the apparatus that produced it malfunctioned. Both Eddington and his critic, the American astronomer W. W. Campbell, followed this rule in interpreting the Brazilian astrographic telescope's photographic plates, each applying their personal plausibility ranking. Eddington thought it very likely that something went wrong with the telescope during the eclipse and so put very little weight on the plates; Campbell thought it somewhat less likely, and so gave the plates more weight. Each used their personal plausibility rankings as proxies for the likelihood of a malfunction; consequently, though they followed exactly the same rule, it instructed one of them to treat the evidence differently from the other.

So it goes with all scientific reasoning: the interpretation of evidence demands likelihoods, and scientists are not only permitted but encouraged to use their subjective plausibility rankings in that role.

When these rankings agree, scientists agree on the treatment of the evidence. In 2016, a marten—a small member of the weasel family—gnawed through a power cable at the Large Hadron Collider at CERN, in Switzerland, destroying itself and seriously damaging the collider's power supply. There was no discrepancy of plausibility rankings in this case: the many scientists working at the facility contemplated the small, smoking corpse and concurred that "something had gone wrong." The collider would need major repairs and subsequent batteries of tests before its data could be trusted. But far more often, plausibility rankings and so the interpretation of data diverge; the subjectivity of the rankings flows directly into scientific reasoning itself. The heart of scientific logic is a human heart.

Kelvin's allegiance to physics over biology and his religiously motivated skepticism about evolution; Eddington's hopes for the new theory of relativity and for international reconciliation after the Great War: did these inclinations derail Kelvin's and Eddington's reasoning, throwing it off the narrow track laid down by objective scientific logic and dragging it into the swamp of human passion and ambition? No; there is no logical track, no one true way, no answer key that science can use to "self-correct" its course. There is only the swamp. Each scientist finds a way through the swamp as best they can. They follow rules, but the rules tolerate and indeed depend on their users' subjective judgments.

Even the darkest and most disturbing demonstrations of scientists' human frailty take on a new hue when plausibility rankings are understood as an essential part of, rather than a corrupter of, scientific reasoning. Scientists sponsored by soda or tobacco companies, I noted earlier, tend to produce results more commercially favorable to those products than scientists with independent funding. Why? The central role of plausibility rankings does allow for cold-blooded calculation: where any of a wide range of rankings could be assigned, a miscreant might intentionally select those that will maximize fame, opportunity, filthy lucre. But although humans are quite capable of such deeds, they are also eminently warm-blooded organisms, whose enthusiasms, hopes, and fears mold their thinking far below the threshold of awareness. Just as referees favor the home team, so scientists' plausibility rankings most likely unconsciously favor their benefactors' businesses. If these were failures to conform to the logically prescribed code of scientific thought, they could be uncovered and corrected by a thorough audit. Examine the rulebook at the points where preferences and prejudice flow into scientists' reasoning, however, and it turns out to say, "Here, apply your plausibility rankings." That is precisely what the scientists in question have done. There is no misdeed; they have acted just as their logic advised.

Science surely does have its malefactors. The rules are sometimes flouted, sometimes deliberately gamed, not least by the leading lights— as on occasion by Newton, Pasteur, Mendel, Haeckel, Millikan, and perhaps Eddington. But even if the advocates of science's powers of self-correction, such as Karl Popper and Atul Gawande, are right to think that such wrongdoing is sporadic or manageable or for some other reason does only limited damage to the scientific enterprise, they cannot appeal to an objective logic to explain science's success. There is no such logic; the evaluation of hypotheses in the light of evidence is thoroughly subjective, fluid all the way down to its core.

THE ESSENTIAL SUBJECTIVITY of the interpretation of evidence is, I have said, not a wholly regrettable thing: by allowing raw, unchecked opinion to animate the process of reasoning, it gives scientific inquiry a vitality and a positive momentum, spurring fruitful argument and competition—Huxley versus Kelvin, Pasteur versus Pouchet—that a more judicious, consensus-making rule would find hard to match.

Yet all the same, by giving up on an objective logic of scientific reasoning, we seem to be abandoning the Great Method Debate, and in so doing losing our grip on the question of what makes science special, of what changed for the better in the course of the Scientific Revolution. It wasn't in the seventeenth century, after all, that the human race first became opinionated or developed a taste for partisan argument.

Objectivity is vital to a methodist such as Popper or Kuhn, you will recall, because it makes possible a systematic, unflagging, uncompromising search through the scientific possibilities, discarding those that exhibit even the slightest weakness. For Popper, what is most important is objectivity's discriminating power: the Popperian rule of falsification purports to detect any discrepancy between theory and the observed facts. For Kuhn, what is most important is objectivity's motivating power. A single set of standards for doing and judging almost everything—the prevailing paradigm—gives scientists the fervent devotion to a research program needed to push it to its empirical breaking point. Neither vision of what distinguishes science from prescientific thinking is viable unless the subjectivity, the individual differences, are sucked out of scientific thought.

Not only through the veins and arteries of all-too-human scientists, however, but also through the hard, rectilinear channels of logic itself, subjectivity surges, giving scientific reasoning its life. By the end of the previous chapter, the notion of an objective scientific method was beaten. It was begging for quarter, blindsided by the self-regard, self-absorption, and slanted perspective of even the sharpest scientific minds. Now it's on its back, discredited, defunct—apparently out of contention for all time.

And yet there is a lifeline: a fine, almost invisible thread of objectivity running through scientific practice. The thread takes the form of a precept regulating scientific argument that is compatible with all I have said about the subjectivity of science so far.

The rule I have in mind allows partialities and power politics to dominate day-to-day scientific inquiry. It allows unfashionable innovators to be ignored and theories with strong social connections to be kept on the books even when their performance is mediocre. But while it tolerates human frailty and condones, indeed draws strength from, the essential subjectivity of scientific reasoning, it brings a subtle pressure to bear. That pressure operates in the long term to do exactly what the methodists have hoped their "scientific method" would do: harvest the facts and distill the scientific truth.

This is the unobtrusive yet irresistible principle that I call the iron rule of explanation. It is time to see it in action.

II

HOW SCIENCE WORKS

CHAPTER 4

The Iron Rule of Explanation

ᘒᘓ

Enter the rule that defines modern science and gives it
unprecedented knowledge-making power.

T o act out the iron rule of explanation, allow me to introduce
two scientists, both alike in dignity, yet separated by character and
culture, education and personal style.

Professor Juliet Capulet, angular and elegant, was raised in school-
rooms, opera houses, and contemporary art museums. She is an afi-
cionado of city life, with its acute organization, its grids and layers of
structure upon structure. An ex-champion in the Multinational Mathematical Gymkhana, she prefers a hypothesis that explains methodically
and exactly, even if it accounts for only a few isolated chapters of the
volume of evidence. For a theory that meets her standards while com-
prehending everything, she is happy to wait.

The energetic and unkempt Professor Romeo Montague grew up
surrounded by grassy expanses and thickets, by birds and beetles. He
likes to think with his hands: having disassembled every household
appliance from the vacuum cleaner to the sump pump, he set his heart
on a life in science to better grasp nature in all its diversity and unruliness. He sees the virtues of mathematical exactitude, but ever restless

and eager to push ahead, he will accept a back-of-the-envelope explanatory sketch with delight if that sketch makes sense of a broad array of phenomena.

Montague cares more about the quantity of evidence explained; Capulet cares more about the higher qualities of the explanation—simplicity, beauty, exactitude. Intensity against extent. Rigor versus reach.

Now let me send these characters back in time to stage some philosophico-scientific theater for your edification and enjoyment. It is based on a true story. The action takes place somewhere in the first half of the nineteenth century, a time during which the nature of heat was the subject of spirited debate. On one side of the debate is caloric theory, created and developed by French scientists from the 1780s through the 1820s, according to which heat is a kind of stuff. This "caloric fluid" flows steadily from warmer to cooler substances, causing them to become hotter and—newly engorged with fluid—to expand. By 1830, caloric theory has achieved a series of spectacular successes: the accurate calculation of the speed of sound through air, the delineation of a mathematical formula capturing precisely the rate with which heat flows through a material such as a metal rod, and an illuminating theory of the efficiency of steam engines and other heat-driven motors.

On the other side is the kinetic theory of heat (sometimes called the mechanical theory of heat), according to which heat is a kind of motion: it is the frenetic, disordered movements of the small particles of which things are made. The theory has a distinguished pedigree—it has been around since the seventeenth century, advocated as we will see by Sir Francis Bacon—but in the light of the French caloric exploits, it is looking rather outworn. Yet there are signs of a revival. The American-born scientist Benjamin Thompson, later to become the Bavarian noble Reichsgraf von Rumford, has just shown that unlimited quantities of heat can be generated by friction, by using a blunt drill to bore a cannon barrel immersed in water. Eventually the water boils—and keeps on boiling as long as the drill is grinding, well after all the system's caloric fluid ought to have drained away. The English chemist Humphrey Davy has conducted similar demonstrations and made similar arguments.

Professors Capulet and Montague find themselves at the center of this standoff, each seeking to explain the behavior of heat as best they can. Capulet's theoretical weapon of choice is the mathematical machinery of caloric theory: systematic, quantitative, precise. Montague, by contrast, is an enthusiast for the kinetic theory: although it cannot yet match the numerical accuracy of the equations of caloric theory, it provides intuitively appealing explanatory mechanisms for a wide range of phenomena, not least the generation of heat by friction.

Who is right: Capulet or Montague? Caloric or kinetic? In 1830, both choices are sound. Twenty years later, the kinetic conception will have overcome the caloric conception. But that has not happened yet. The choice between the two is dictated not so much by evidence as by temperament and taste. And so Montague and Capulet eye one another uneasily from opposite sides of the ideological stage.

In a few moments, the curtain will rise. That leaves just enough time for a brief authorial disclaimer: what follows is fiction, not historical fact. The background is historically real, as is the important problem with which the play begins. After that it is for the most part philosophical embellishment, intended not to tell the story of the science of heat but to direct the spotlight past the players and onto the backdrop, exposing a critically important, fixed plane of scientific agreement that lies behind the essentially subjective scientific dispute concerning which theory to believe.

Very well; let the drama begin. Capulet and Montague are deep in argument. The question is heat's ability to radiate across empty space, as when the sun pours its life-giving warmth across the interplanetary vacuum onto the earth. Such a thing seems impossible if the kinetic theory is correct, as there is nothing in a vacuum to vibrate, so nothing that can transfer the tiny motions that make up heat from source to destination. Perhaps Professor Capulet is right to reject kinetic theory, then—for surely here it finds a definitive refutation?

An imaginative Professor Montague wonders, however, whether heat might not travel through empty space in a different form than it travels through air, earth, and water. Perhaps, he suggests, there is a process—call it radiation—by which the vibratory energy of a hot body is transformed into "heat rays" that, like light rays, fly through empty space at high speed and, colliding with solid matter on the other side, induce the sort of vibrations that amount to heat proper? Professor Capulet considers that to be a transparently desperate attempt to save the kinetic theory from what is to any neutral observer an obvious falsification.

Montague's saving maneuver is, however, not only logically permissible, but factually quite correct. The sun's heat arrives at the beach thanks to electromagnetic radiation of certain frequencies, much of it in the infrared frequency band. These heat rays are not themselves a form of heat; rather, heat is transformed into electromagnetic energy, which travels as a ray and is then, upon striking the sand and sundry bathers' bodies, changed back to heat, exactly as Professor Montague's convoluted explanatory narrative maintains.

That said, Capulet's skepticism has a sound rationale. She needs only one kind of thing, caloric fluid, to explain the movement and behavior of heat. Montague has posited two kinds of things, heat itself and heat radiation, each popping up to do the job it is good for and then conveniently metamorphosing into the other when that's what's needed instead. At the very least, Montague seems to be violating the old methodological dictum "not to posit two entities when one will do"—usually, in honor of the medieval philosopher William of Occam, called "Occam's razor." It is hardly unreasonable for Capulet to threaten Montague with a close shave.

This is one more example, then, of the essential subjectivity of scientific reasoning: the wisdom of Montague's willingness to postulate radiation—a whole new way that heat might be shipped around the universe—depends on your perspective. To Montague, it is an inspired discovery; to Capulet, poetic indulgence. One scientist senses the first glimmer of the sun's rays and the lark's joyful song; the other only darkness and the plaintive anthem of the nightingale.

Montague and Capulet disagree as to which theory best explains the evidence. And they disagree on how particular pieces of evidence —most notably, heat's ability to travel through a vacuum—bear on each other's theories. Is this the version of *Romeo and Juliet* in which the lovers survive, only to have family squabbles lead slowly but inexorably to irreconcilable differences?

We expect Elizabethan drama to end stirringly in true love or violence, and modern drama (perhaps) in misery or self-recognition. The drama of science, however, never need end at all. Montague and Capulet can continue their dialogue indefinitely without running out of lines. That is because there is always a purely scientific way to perpetuate a scientific argument: make more measurements or conduct new experiments.

The unending script, the code of conduct for scientific argument according to which Montague and Capulet may continue their debate indefinitely, is provided by the methodological precept that I call the iron rule of explanation. What the rule says is simple enough: it directs scientists to resolve their differences of opinion by conducting empirical tests rather than by shouting or fighting or philosophizing or moralizing or marrying or calling on a higher power. That is all; it makes no attempt to interpret the evidence, to decide winners and losers. Indeed, its function is not so much to resolve the dispute as to prolong it. This perpetuation of the dramatic conflict for its own sake is the essence of the scientific method, and as the perpetuator-in-chief, the iron rule establishes itself as the heart, the soul, the life force of scientific inquiry.

The first act has ended in impasse. Montague said that heat traveling across empty space takes the form of radiation; Capulet said that it is a spume of caloric fluid jetting through the void. The iron rule shows them how to go on arguing, how to build a second scientific act around their disagreement, by improvising a series of experimental scenes—that is, by conducting a succession of empirical tests.

To decide the question of how heat moves through the void, Montague and Capulet might, for example, agree to measure the difference that a pane of glass makes to the velocity with which heat travels across a laboratory-made vacuum. If Montague is right and heat is transmitted by radiation, then heat rays should speed through the glass like light rays, barely slowing at all. If Capulet is right and heat is an invisible, flowing liquid—caloric fluid—then the glass should stop the heat, or at least slow it down significantly, as it works its way by diffusion from one side of the pane to the other.

Suppose they perform the experiment. The glass pane makes no difference to the time taken for heat to traverse the vacuum chamber, so the test looks to count in favor of Montague's theory and against Capulet's. Capulet might surrender, but then again she might fight on. After all, her caloric hypothesis predicts a significant slowdown only in conjunction with certain auxiliary assumptions—in particular, the assumption that the glass pane acts as an effective barrier to caloric fluid. Perhaps that's not the way it works. Perhaps firing heat at a pane of glass is like directing a fire hose at a fishing net: the stuff courses through holes invisible to us but huge by comparison with the particles that make up the fluid (if it is made of particles at all). To test Capulet's "fishnet" response, Montague concedes, more tests will be needed. Perhaps stack up many panes of glass at different angles? If the fishnet analogy holds, sufficiently many layers should slow caloric fluid right down.

Or suppose that the glass pane experiment comes out the other way: the glass slows heat flying down the chamber to a crawl. Exit Montague?

Figure 4.1. Montague proposes an experiment.

Not necessarily. Glass is transparent to the radiation that we call light, but perhaps it is opaque to heat radiation, acting like a thick black curtain. To test this "blackout" response, Capulet concedes, more tests will be needed. Perhaps try various other potential barriers? If heat rays are like light rays, then in the same way that some materials are transparent to light, some ought to be transparent to heat.

Whatever the outcome of the test, then, two things are certain. First, the "loser" will have the chance to save their theory by rejecting one auxiliary assumption in favor of another. Second, both scientists will agree on the kinds of further observation that will put the new auxiliary assumptions to the empirical test. Whichever way the glass pane experiment goes, for example, Capulet and Montague might agree to pursue their argument by trying different materials as barriers in the vacuum chamber to find out what lets heat through and what does not.

Because they follow the iron rule, Capulet and Montague are in a certain kind of agreement. It is not agreement as to the best theory of heat. It is not agreement as to what the evidence says about various theories of heat. It is rather a kind of procedural agreement, an accord as to how to go on arguing: by observation and experiment, and not in any other way.

Simply saying, "Resolve your disputes through empirical testing" is not enough to establish an accord, however; what's also needed is a shared sense of what counts as an empirical test. The iron rule provides the necessary definition, an objective criterion for empirical testing to which all scientists subscribe.

You have two hypotheses that you want to decide between. Is heat a special kind of substance, different from ordinary matter, or is it the disordered motion of small particles of matter? Is the earth 20 million years old or more than 100 million years old? Do its continents move, or do they stay in place? To answer these questions, the iron rule says, proceed as follows: find an experiment or observation that might have one of two possible outcomes, where the first outcome would be explained by the first hypothesis (or rather, a cohort containing the first hypothesis) but not the second, and vice versa. Perform the experiment or make the observation. See what happens.

Here, then, in short, is the iron rule:

1. Strive to settle all arguments by empirical testing.
2. To conduct an empirical test to decide between a pair of hypotheses, perform an experiment or measurement, one of whose possible outcomes can be explained by one hypothesis (and accompanying cohort) but not the other.

There lies the nub of the scientific method and so, once its subtleties have been spelled out in the chapters to come, the denouement of the Great Method Debate.

THE IRON RULE GIVES all scientists the same advice as to what counts as a relevant experiment or observation, regardless of their intellectual predilections, cultural biases, or narrow ambitions. It does not, however, pretend to prescribe what to believe on the basis of such tests. It is a rule for doing rather than thinking.

This uniformity of doing, this procedural consensus, may not sound like much. It is an etiquette for argument, an agreement on how to disagree. What can it bring to science, beyond a certain civility and orderliness? In spite of its apparently modest ambitions, the procedural consensus is precisely what secures the triumph of modern science.

The first benefit of the procedural consensus prescribed by the iron rule is simple continuity. Throughout history, religious traditions have been vulnerable to schism—to irreparable ideological separations yielding daughter traditions that lose the ability to reason with one another. In their mutual future lies suspicion, political jockeying, sometimes appalling violence. Islam split into its Sunni and Shiite branches; early Christianity into Roman Catholic and Eastern Orthodox camps; Roman Catholicism into numerous Protestant factions that denounced and battled the papal loyalists.

Philosophical and political traditions have the same susceptibility. Even natural philosophy, the study of nature that is the precursor of modern science, has at times fractured into factions that, though they coexisted peacefully, found little to talk about. Physics after Aristotle bifurcated into two schools, the Epicureans and the Stoics, the one affirming an atomistic view in which the universe is composed of nothing more than particles careening blindly through the emptiness of space, the other a view in which matter fills the universe according to the dictates of rationality. An individual seeking enlightenment while Stoicism and Epicureanism held sway, from about 300 BCE to 300 CE, could choose one school or the other or attempt to learn from both, but the schools themselves remained detached intellectual traditions, separated by both their physics and their philosophy of life, until economic decline, barbarian invasions, and Christianity finished them off.

Since the creation of modern science in the seventeenth century, empirical inquiry has seen nothing like this intellectual fission. There was no irreconcilable split between the caloric and kinetic schools of heat, nor did Kelvin's work cause biologists to cut off all conversation with physicists. They continued to work together because in the iron rule they had a shared rule of engagement and no choice but to use it.

Peaceful dialogue is a start, certainly better than the alternative, but it is hardly sufficient to guarantee progress, let alone science's fabled power to zero in on the truth. There must be more. The clues lie in the pioneering work of Popper and Kuhn on the importance of motivation and morale in science.

Scientific argument, unlike most other forms of disputation, has a valuable by-product, and that by-product is data. The iron rule encourages, instructs, obliges, or forces contending scientists to engage each other with observable fact alone. Theological and philosophical contentions are ruled out of court; as long as the protagonists are practicing science, their arguments must be conducted by experimental means. All of their need to win, their determination to come out on top—all of that raw human ambition that, on the modern sociological view of science, would subvert any objective code of inquiry—is diverted into the performance of empirical tests. The rule thereby harnesses the oldest emotions to drive the extraordinary attention to process and detail that makes science the supreme discriminator and destroyer of false ideas.

I wrote earlier about the long, tedious years of brain crushing and distillation required for the rival scientists Roger Guillemin and Andrew Schally to analyze the structure of the hormone TRH. (Schally estimated that in the course of his efforts to find the structure of another hormone, called LRF at the time, he had to process the hypothalami of 160,000 pigs to obtain less than a thousandth of a gram of the sought-after substance.) What kept them going? In part, as Schally pointed out, the Kuhnian belief that substances such as TRH and LRF existed and that current techniques were capable of revealing their structure. Everyone had that belief, however. The thing that made the difference between the Nobel Prize winners Guillemin and Schally, on the one hand, and their rivals who failed to complete the project, on the other, was the lust for victory. As Schally, who was born in 1926 in Vilnius, then part of Poland, said of one of those lesser rivals:

> He is the Establishment . . . he never had to do anything . . . everything was given to him . . . of course, he missed the boat, he never dared putting in what was required: brute force. Guillemin and I, we are immigrants, obscure little doctors, we fought our way to the top.

That extraordinary combativeness could have manifested itself in fierce metaphysical argument or in fine rhetoric. But because Schally's enterprise was bound by the iron rule, his ardent spirit was directed exclusively toward the production and analysis of TRH and so toward the generation of empirical data.

There is still more to the motivational power of the iron rule's procedural consensus. Think about chess. The rules of the game are simple and known to all. No player wastes time musing on or debating the rules; they simply follow them. Consequently, enormous energy is liberated to pursue the problem how best to play within the established regulations.

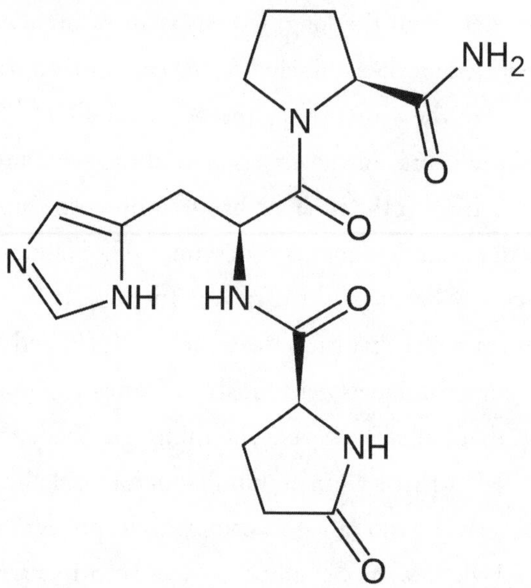

Figure 4.2. The fruits of procedural consensus: the molecular structure of TRH.

There are numerous compilations of openings, analyses of midgame strategies, commentaries on individual contests. No one doubts that this effort is important (to those who care about chess), because no one questions the framework within which it is carried out. Following the rules is just what chess is.

The iron rule establishes certain "rules of the game" in a similar way and with similar effect. The game is scientific argument, and the iron rule specifies what counts as a "legitimate move"—namely, performing an experiment or making an observation that generates relevant empirical evidence. Because the rule imposes a standard for what counts as empirical testing that never changes, questions of legitimacy are settled once and for all. Scientists' minds and money are thereby freed to focus on what can be done within the prescribed framework and thus on the question not of what makes for a legal move but of what makes for a good or excellent or brilliant move.

The consensus on testing enables, for example, an across-the-board agreement that certain experiments ought to be performed, even experiments that consume disconcertingly large quantities of labor and capital. In some cases, scientists who may have quite opposing theoretical perspectives and ambitions come together to form a temporary community sharing an experimental purpose and capable, because of its size and unity, of doing things beyond the means of a single research group. Of the scientists who searched for the top quark at Fermilab near Chicago in the 1990s or the Higgs boson at the Large Hadron Collider near Geneva in the 2010s, some were hoping to confirm the predictions of physics' Standard Model—which implies that these particles exist—while others were hoping to overturn the Standard Model ("I *hate* the Standard Model," said one to me), and others still were agnostic. They agreed, nevertheless, that the experiments were worth all their time and a lot of money.

More typically in science, research groups undertake their experimental projects independently, as in the case of Eddington's expedition to see the eclipse or Guillemin and Schally's independent attempts to isolate TRH. They set off alone, but thanks to the consensus brokered by the iron rule, they do so with the approval of their competitors, who agree that the experiment or observation will make a valid empirical contribution to the argument. This approval constitutes a kind of moral support, and in the modern era, where science is conducted mostly with other people's money, it also often enough secures financial support, as boards consisting of scientists with opposing views agree to direct funding to what seem to all members to be worthwhile projects—even if they hope for contrary outcomes.

The game of science is not always played against direct rivals. Sometimes it is more like a roomful of solitaire players, each striving to outdo the others in a competition to solve the multifarious problems posed by nature. The contest makes sense, though, only if everyone strives under the same unchanging regulatory regime.

And so, spurred on but at the same time boxed in by the iron rule, scientists conduct new experiments or make new observations that no one would otherwise have had the motivation or the resources to bring about—using satellite antennae to pick up patterns in the faint radio echoes of the Big Bang, gravimeters to detect minute differences in the thickness of the earth's crust, calorimeters to measure impalpable flows of energy in chemical reactions, gentle yet comprehensive techniques of excavation to reveal fossilized feathers, shattered pots, fallen temples—each carefully calibrated to favor, on empirical grounds only, one theoretical cohort over another.

My inspiration for emphasizing the importance of regulatory unity, I hope you see, is Kuhn's conception of paradigm-driven science. I have my differences with Kuhn. His games are individual scientific research programs—Newtonian physics, say, or evolutionary biology—each of which has its own distinctive set of rules, implicit in the paradigm, determining the permissible moves within the program. My game spans all of science: its iron rule governs Newtonian physics, microeconomics, and molecular genetics alike. For Kuhn, then, the rules change with the times—revolutionary times, that is—whereas like Popper's rule of falsification, my procedural consensus is forever, or at least for as long as science endures.

Further, a Kuhnian rule-making paradigm is a psychological idée fixe, alternatives to which cannot even be conceived by the scientists within its grip. I hold, by contrast, that science's iron rule is more like a sporting convention than a mental prison. Chess players can easily imagine different rules (and occasionally entertain themselves by doing so), but for the purposes of the competition, they agree to stick with the official set. Likewise, scientists abide by the iron rule not because it is impossible for them to conceive of an alternative, but because they know that the rule characterizes what it is to do science, and just as chess players want to play chess, they want to do science.

Although my disagreements with Kuhn are deep and important, I emphasize the extent to which I have taken over Kuhn's fundamental insight: that establishing beyond question the legitimacy of evidential maneuvers, whether by paradigm or by the unchanging iron rule, provides an intellectual security that supercharges the scientific enterprise's power to generate expensive but telling empirical tests.

The ideas in the last few pages are some of the most important that I have to share. Together they capture the way in which the procedural consensus orchestrated by the iron rule powers the scientific knowledge machine. I'll state them one more time.

First, the consensus ensures continuity. Provided that there are resources and the will to continue the investigation, there is no prospect that Montague and Capulet will arrive at a position where they can think of nothing to do or say that might bring them any closer together. There are no duels or divorces in science, no schisms to parallel the historical schisms in religion, politics, and philosophy, where opposing camps stop talking or worse. Always there is something that even the most bitter adversaries can agree to do next: another test.

Second (and consequently), the iron rule channels hope, anger, envy, ambition, resentment—all the fires fuming in the human heart—to one end: the production of empirical evidence. Capulet and Montague's complex feelings about one another must, as long as they follow a scientific script, work themselves out in the lab, the field, or the observatory.

Third, the fundamental conventions of the game, the legitimate evidential moves, are fixed for all time. This gives the players, the scientists, a quiet but firm certainty in the foundation of their enterprise, a stable intellectual and moral platform on which to construct great experimental enterprises, with all their financial, emotional, and physical demands.

In these ways, a protocol as weak as the iron rule—an agreement to argue by empirical testing alone conjoined with a fixed standard to determine what counts as an empirical test—gives science something that no other form of inquiry before it has had, a pool of empirical observations that dwarfs in its size, scope, subtlety, and precision anything the ancient or medieval natural philosophers could bring themselves to produce.

THE GREAT RESERVOIR of evidence tapped by the iron rule is essential fuel for science, but can it drive scientific progress unaided? Does a great mass of empirical data, left to stew, convert itself spontaneously into proportionally great quantities of theoretical truth?

It is not as easy as that. Empirical knowledge cannot be transformed directly into theoretical knowledge. As Eddington's telescopes, Kelvin's estimates of the earth's antiquity, and Montague and Capulet's measurements of heat radiation show, observations must be carefully interpreted if they are to tell us what theories to believe. There lies a daunting problem. I have given up on the objectivity of scientific reasoning and proposed in its place a purely procedural agreement. The agreement may bring forth gratifyingly large quantities of data. But without an objective scheme of interpretation, data in any quantity is useless. It says a different thing to each scientist.

How, then, does science go about finding the truth? A promising idea lies at the juncture of Shakespeare's death and the dawn of the Scientific Revolution. Let us go there.

CHAPTER 5

Baconian Convergence

෨෬

How science's consensus on procedure, enforced by
the iron rule, leads to discovery

I N 1618, Sir Francis Bacon was elevated to the post of Lord Chancellor of England, becoming one of the highest legal officials in the land and a close adviser to the king, James I. Just three years later in 1621, Bacon was denounced by his political enemies and put on trial for bribery. He escaped with his life but not his honor, retiring in disgrace. Somehow, in the middle of that short, tumultuous time, Bacon published one of the most significant books ever written on scientific inquiry: *The New Organon*, a blueprint for the knowledge machine that was to be constructed over the subsequent decades of the Scientific Revolution. Such was the book's fame that the poet Abraham Cowley, in a paean that prefaced Thomas Sprat's 1667 history of the Royal Society, declared Bacon to be "Lord Chancellor of the laws of nature."

Bacon admired the ancient Greek natural philosophers enormously, yet they had plainly failed, he saw, in their endeavors, their inquiries going "round in circles for ever, with meager, almost negligible, progress." Thus, "a new beginning has to be made from the most basic foundations": the old philosophical ways would have to give way to a novel method for discerning the deep structure of the natural world.

The primary casualty of Bacon's war on the past was to be Aristotle's *Organon*, meaning "the tools," a treatise on the rules of reasoning and the proper organization of knowledge regarded as the supreme work on these topics during the Middle Ages. It would be replaced, Bacon hoped, by his own work, whose ambitions were made clear on the title page: his *New Organon* would supersede Aristotle's ancient work.

When the book was published in 1620, the Scientific Revolution was barely underway. Galileo had used a telescope to observe the moons of Jupiter in 1609, calling into question the Aristotelian thesis that all celestial bodies orbit the center of the universe (which Jupiter most certainly was not). In the same year, Kepler published his first two laws of planetary motion, describing in mathematical terms the elliptical orbits of the planets around the sun. But the *New Organon* is not so much a part of the Scientific Revolution's first act as it is its Shakespearean prologue, laying the groundwork, heralding the action, anticipating the denouement, from the viewpoint of one who sees all that is to come—the playwright himself.

A method for discovering everything—that was what Bacon promised his readers. To show off its power, he proposed to apply it to a great mystery of the time, the nature of heat (the very same topic debated in the previous chapter by the Baconian characters Montague and Capulet).

The natural philosophers before him had failed to solve the problem of heat, as they had failed in so much else, but it was not, Bacon thought, for lack of clues. To understand nature, you must open your eyes to these clues. But before you begin, you must empty your mind.

You must put aside your personal prejudices, loyalties, and preferences. You must put aside the prejudices common to the human race, such as the tendency to oversimplify, to neglect to look for evidence against your pet theories, or to pay more attention to freakish than to everyday phenomena. You must put aside all prejudices inherent in your language—you must not assume, for example, that two things having the same name have a similar nature. (Just because a hairball is made of hair, don't expect a football to be made of feet.) You must put aside all philosophy and therefore any prior science that is based even in part on philosophical thinking, such as the science of the ancient Greeks. Indeed, you must put aside all learning that is not produced by the Baconian method. These distractions and temptations Bacon called "idols"; to allow them to influence your thinking was to worship false gods rather than to revere reason.

Can any human being hope to discard not only their particular idiosyncrasies, but also the suppositions and inclinations that are their birthright as an educated person, as a member of their culture and of their species? When at the height of his brilliant career Bacon was put on trial for corruption, he declared that he was "as innocent as any born upon St. Innocents Day"—because he had not allowed the bribes he had taken to sway his judgment. Perhaps he did indeed have a mind strong enough and a soul stern enough to resist any urge to favor those who had smiled on him, but the stories of Eddington and the eclipse, of Pasteur and spontaneous generation, of continental drift and the age of the earth suggest that scientists, even great scientists, are rather more pliable than that.

Never mind. Let's suppose that Bacon's mind is successfully purged of all inclination to venerate the idols; let's see how he pulls together the clues to infer the nature of heat.

His first step: assemble all positive instances of heat, that is, all types of circumstances in which heat is present: in fire, in the bodies of animals, when the rays of the sun are concentrated using a magnifying glass, when two solid bodies are rubbed together. And then there are flaming meteors, natural hot baths, spicy herbs, and more.

His second step: assemble all negative instances of heat, that is, all types of circumstances in which heat is not present. Of course, this list must be endless, so Bacon recommends the following alternative. For each of the positive instances, find similar circumstances in which heat is absent. For example: heat is absent in the bodies of dead animals or when two solid bodies are held together but not rubbed (and apparently in moonbeams, comets, and St. Elmo's fire—though as Bacon remarks, further investigation is needed).

The third and final step is to assemble all the ways in which heat varies with other quantities, for example, the way that objects get hotter the closer they are to a fire or the way that metals take longer to get hot than air but retain their heat longer.

With the third step completed, Bacon has the evidence he needs before him. What, he asks himself, can possibly explain all the assembled facts? What can explain why heat is present in live animals but not dead animals, why it is generated by friction, why metals soak up more heat than air? He considers many suggestions concerning the essential character of heat. Each explains at least a few of the positive and negative instances but fails to explain others and so is rejected: "Every contradictory instance destroys a conjecture about a form." Heat cannot be "light and brightness," because substances such as boiling water can be hot without being light or bright. Heat cannot be a substance, because when a hot iron warms another object, it loses heat but does not lose any weight. And so on.

At the end of the process, there is only one hypothesis standing, only one conjecture about the nature of heat that can explain every circumstance upon which Bacon has trained his gaze. It is the hypothesis that explains how friction generates heat: "The quiddity of heat is motion and nothing else," or, more exactly, heat is the disordered motion—the vibration—of the small particles of which all things are made. This idea, an early version of the kinetic theory, is, we now know, quite correct: a compelling advertisement for the Baconian method.

The *New Organon* recommends this same technique to investigate every natural phenomenon, from lightning to laryngitis to life itself: gather the conditions under which the phenomenon occurs, the conditions under which it does not occur, its patterns of change, and find the hypothesis that explains the lot—the occurrences, the nonoccurrences, the variation. That hypothesis is the theory you're looking for, the truth.

Bacon imagined that once scientists got to work using his discovery method to understand a phenomenon, they would quickly rule out all explanations but the correct one. That turned out to be rather optimistic. In the hundred years or so after Bacon laid down his prescriptions, the number of interesting, plausible explanations was multiplying rather than declining, as the great creative minds of the seventeenth and eighteenth centuries went to work attempting to make sense of the world. Bacon's triumphant revelation of the kinetic nature of heat, for example, was (as you know) challenged around 1800 by the development of the rival caloric theory, which was thought at that time to better account for the assembled facts. To scientists facing Capulet and Montague's predicament, Bacon's method provides no objective guidance in determining the nature of heat. They must, as we saw, fall back on their personal estimates of the likelihoods of various possibilities—their plausibility rankings—which inevitably means letting the idols, Bacon's icons of subjectivity, have their say.

The point of our visit to the early seventeenth century was not, however, to understand why copies of the *New Organon* are rarely spotted on the modern laboratory workbench. It was to understand how science eventually gets to the correct view of things. The difficulties in using the Baconian method arise not because there is something wrong in principle with Bacon's assumption that the truth and only the truth can explain everything, but because it is almost impossible to apply in practice: in the middle of scientific inquiry, there are typically several competing hypotheses of roughly comparable explanatory success.

At some point, nevertheless, the conflict begins to subside. The moment that Bacon longed for arrives: the treasury of evidence has become so rich that it singles out just one theory as supreme explainer, the sole theory capable of making sense of every observable pattern—in Bacon's terms, every "positive instance," every "negative instance," every pattern of variation. Viewed from this cosmic perspective, Bacon's idea looks not so different from Popper's. Each false theory has been confronted with some piece of evidence that it cannot explain, and in the light of its explanatory inadequacy has been tossed aside. Only the truth remains.

Popper rejected any role for plausibility rankings in scientific reasoning; they were, to his philosophical taste, contemptibly inductive. Let them back in, though, and you'll find that something interesting happens as science closes in on the last theory standing. As evidence accumulates, plausibility rankings begin to converge. Differences in opinion become less extreme. Consensus emerges as to which are the leading theoretical contenders and which are the also-rans, then eventually on which is the best of them all. There is not complete agreement, but there is ever less disagreement. This is *Baconian convergence.*

We do not have to wait indefinitely for Baconian convergence to do its work. The knowledge machine has, in just a few hundred years, discovered viruses, DNA, the nature of heat, the genes underlying most animals' basic body plans, the family relationship between humans and chimpanzees, the family relationship between English and Hindi, undersea volcanoes, tectonic plates, the moons of Jupiter, the rings of Saturn, the Andromeda galaxy, black holes, the atomic structure of gold, the difference between carbon and diamond, the function of the heart, and the architecture of neurons. In each case, the weight of observable fact came to overpower differences in plausibility rankings. Contrary opinion became quirky, then eccentric, then laughable. Evidently, Baconian convergence is real and well within scientists' reach.

Some circumspection is required. It is not always clear when convergence is complete. For a long time it seemed that Newton had gotten gravity basically right; then Einstein came along to upend that complacency. A new quantum theory of gravity might yet show that even Einstein did not have the full story. And science's truthward trajectory is often not a steady march. In the course of finding correct explanations, opinion may temporarily move away from literal truth, as the science of heat moved in the nineteenth century from the kinetic to the caloric theory. Baconian convergence is, in the short or even the medium term, a fitful process.

Furthermore, the observable facts themselves are not quite the bedrock that Bacon takes them to be: published evidence can be "bad" in various ways. Often, due to some problem with the instruments, it is misleading. The plates from Eddington's Brazilian astrographic telescope, for example, wrongly suggest a Newtonian gravitational bending angle for light. Sometimes evidence might even be manipulated, as historians have suspected of "facts" reported by Mendel, Haeckel, Millikan, and Newton. If Baconian convergence is to occur, science must somehow neutralize bad data.

Theorists who say that science is "self-correcting"—I quoted both Karl Popper and Atul Gawande to this effect in Chapter 2—believe that when it really matters, bad data is eventually recognized as such. In a sense, that is indeed what happened with the photographs from the Brazilian astrographic: now that we know that Einstein was right (or at least, more right than Newton), we can look back and say that the astrographic setup must have been faulty. But such reasoning is possible only because the scientific community has already converged on and endorsed general relativity. Corrections of this sort cannot, therefore, explain how convergence occurs in the first place.

What happened in the case of general relativity, and what tends to happen in science more generally, is that opinions converge not because bad data is corrected but because it is swamped. After Eddington, many determinations of the gravitational bending angle were made and many other tests of relativity conducted. The great preponderance of the resulting measurements fit Einsteinian physics better than Newtonian physics. Eddington's data, right or wrong, simply ceased to matter.

So we see that over the decades and centuries, agreement emerges again and again from the tissue of uncertainty and dissent that is the characteristic stuff of science—provided that everyone keeps arguing and the stock of observations continues to grow.

Written by Sir Francis Bacon, *Romeo and Juliet* would have had a happy ending. For all their rivalries, their contrasting temperaments, their contrary theoretical tastes, their disparate explanatory standards, Professors Montague and Capulet would be pressed together, eventually, by sheer weight of evidence. Their dialogue, strictly schooled by science's iron rule, would line by line, fact by fact, bend their wills to the world and so bring their minds at last into perfect congruence.

TWO PRINCIPAL PARTS of my explanation of science's power, of my contribution to the Great Method Debate, are now in place: the iron rule's procedural consensus and the Baconian convergence that the consensus makes possible. There is one further element to the story.

In a basement in Ohio in 1887, Albert Michelson and Edward Morley sent two beams of light flying. One beam was traveling in the same direction as the earth's motion around the sun; the other was traveling at right angles to the first. After traversing identical distances, the beams were reflected back to their source and compared. If their waveforms, when superimposed, failed to overlap exactly, it was because one was taking longer to make its journey than the other—

which is precisely what Michelson and Morley expected. The aim was to measure the speed with which the solar system was traveling through the "ether," an invisible substance hypothesized by nineteenth-century physicists to be the bearer of light waves in the same way that water is the bearer of ocean waves and air is the bearer of sound waves. The faster the movement through the ether, the slower the total travel time of the beam traveling in line with the earth, so the greater the discrepancy between the two superimposed beams. One momentous consequence of such a measurement would be direct evidence, at last, for the ether's existence.

It was a finicky business. The distance by which one beam would be shifted relative to the other was expected to be on the order of a few hundred nanometers, or about 0.00001 inches. The smallest vibrations—passing horses or distant claps of thunder—could upset the measurement device. That is why it was situated in a basement, mounted on a huge sandstone block, floating in a trough of mercury (Figure 5.1). That is also, perhaps, why Michelson had, partway through the 10 years

Figure 5.1. The apparatus that Michelson and Morley used to measure the speed of light in various directions relative to the earth's direction of travel.

he devoted to building ever more sensitive mechanisms to determine light's speed, suffered a nervous breakdown—so conjectured his collaborator Morley.

The end result was, as Michelson put it, "decidedly negative": the experiment had failed to detect any motion at all with respect to the ether. For a time, there was contention and confusion about both ether and the experimental setup. It was cured by Einstein, whose special theory of relativity dispensed with the ether and explained exactly what Michelson and Morley had unwittingly observed: that the velocity of incoming light is the same regardless of whether you are stationary with respect to the light source or traveling toward it at high speed. After two hundred glorious years, Newtonian physics was dead. That is the difference made by 0.00001 of an inch.

As you have already seen, Einstein's treatment of gravity was tested a few years later by scrutinizing similarly tiny details: Eddington's eclipse experiment measured a shift in the apparent positions of stars equal to about one-third of the apparent diameter of Mars at its smallest. The Gravity Probe B satellite, also testing general relativity, was designed to pick up discrepancies of about 0.00001 degrees per year. Newtonian and Einsteinian physics tell vastly different stories about the structure of the universe. But to see which of them is correct, you need to make almost infinitesimally small measurements.

That the secrets of the universe lie in minute structures, in nearly indiscernible details, in patterns that only the most sensitive, fragile, and expensive instruments can detect, is an insight so important that it deserves a name. I call it the *Tychonic principle*, after the sixteenth-century Danish astronomer Tycho Brahe, who, working just before the invention of the telescope, was the last and greatest "naked-eye" astronomer, using sextants and quadrants to pinpoint the positions of stars and the movements of planets to within 0.02 degrees. To achieve this level of accuracy, Tycho built an observatory called Stjerneborg ("castle of the

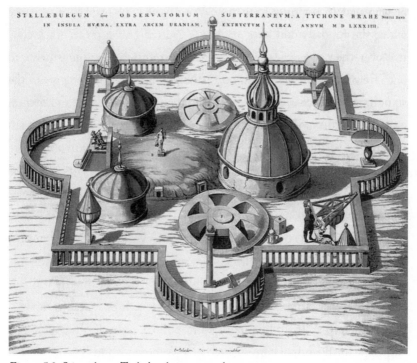

Figure 5.2. Stjerneborg, Tycho's subterranean observatory.

stars") entirely at basement level, seeking as Michelson and Morley later would to shelter from the inaccuracy inflicted by the bustle and noise of the world outside (Figure 5.2).

The Tychonic principle holds not only in fundamental physics and not only for strictly quantitative detail. To discover the intricacies of biological heredity and evolution or the way that multicellular organisms develop and grow from embryos, researchers must attend painstakingly to complex causal structures at the microscopic level. Similar attention to exact amounts and elaborate connections is required of neuroscientists, geologists, and archaeologists, of model builders in climate science and economics, and in a somewhat different way of anthropologists and sociologists.

Let me test your imagination with an outré thought: the Tychonic principle need not have been true. We might have lived in a universe where the fundamental laws of physics exerted only a rather slack control over the trajectories of particles, determining their movements to one or two decimal places but no further. Making measurements to the sixth decimal place in such a world would be a waste of time and money, no more conducive to knowledge than counting pebbles on the beach or drops of water in the sea. Equally, we might have lived in a universe where the forms of organisms were determined by vital spirits, or by the whimsy of the gods, rather than by intricately structured chains of molecules. But we don't; our world, it seems, is Tychonic through and through.

Bacon proposes that if we take explanatory power to be our guide, then empirical testing will ultimately single out the truth. The Tychonic principle says that much of that testing will be extraordinarily difficult to conduct. It is for this reason that the iron rule is essential to modern science's success. The effort required to build a store of observable fact sufficient for Baconian convergence in a Tychonic world is so great that humans can be persuaded, induced, or impelled to take on the project only under exceptional circumstances. The iron rule engineers those circumstances and so "forces scientists to investigate some part of nature in a detail and depth that would otherwise be unimaginable." Those are Thomas Kuhn's words. He attributed the compulsion to paradigms, not to a universal procedural consensus—otherwise, this was the best sentence he ever wrote.

THE IRON RULE tells scientists to pursue the truth by looking for the theory that best explains the observable facts. What is so groundbreaking, so revolutionary about that? Very little, some might say. We humans have known how to infer from what we see to what best explains it since our Paleolithic ancestors first grasped the significance of saber-tooth footprints in the snow.

Further, the value of this explanatory way of thinking was as obvious to ancient and medieval thinkers as it was to any prehistoric hunter. The first Greek philosophers—Thales, with his theory that everything is made of water; Heraclitus, who preferred fire; Anaximander, who thought that the fundamental constituent of things is the "boundless"— were all doing their best to account for what they saw in the world around them: rainbows, magnetism, fossils, epilepsy, the saltiness of the sea, the starriness of the sky, the exquisite adaptedness of living things to their natural habitats. Aristotle single-handedly added a dazzling array of phenomena to this list, from tornadoes to respiration to human hands and waterbirds' webbed feet, and offered systematic theories of physics, biology, and psychology to explain them. Indeed, he is sometimes held up as the first great scientist in history.

The eleventh-century Islamic philosopher Ibn Sīnā, called Avicenna in the West, laid down seven rules for medical experimentation. The English scholastic philosopher Robert Grosseteste developed the notion of a controlled experiment in the high Middle Ages, centuries before Bacon. And experiment was not only praised in principle but deployed in practice. The historian David Lindberg provides a partial list of experimenters at work before the Scientific Revolution that stretches from the ancient Roman Empire to Islamic Persia through the European Middle Ages, including, among many others, Ptolemy, Ibn al-Haytham (Alhazen), Kamāl al-Dīn, Theodoric of Freiberg, Rabbi Levi ben Gershon, Johannes de Muris, and Paul of Taranto.

What does the iron rule add to all this? Where is its novelty? On what grounds can I say that Aristotle was not a modern scientist when he manifestly valued the power of his theories to explain the phenomena that he so keenly observed?

For all its emphasis on explaining the world and its openness to experimentation, the natural philosophy of old could not establish the procedural consensus responsible for modern science's superlative knowledge-making power. To formulate the iron rule and thus to make modern science, something had to be added to the age-old doctrine that true theories can be recognized by their explanatory power. That supplement takes the form of four methodological innovations.

The first innovation is a reformulation of the material that would make up the iron rule—explanation itself. Before the Scientific Revolution, explanation was mixed with philosophical principles, and so explanatory power was a subjective matter, varying with the beholder's intellectual commitments and temperament. In modern science, the notion of explanation is free of philosophy and all other ideology; it is as pure as the elemental metal from which the iron rule takes its name. Consequently, explanatory power means the same thing to every modern scientist, regardless of upbringing and inclination, so that every scientist agrees on what the iron rule says, on what satisfies the rule's criterion for a meaningful scientific test.

The second innovation is a matter of place, of venue, of domain. The iron rule is focused not on what scientists think, like the traditional laws of logic formulated by Aristotle, but on what arguments they can make in their official communications. Their brains are left unfettered by the rule, then; it is their public pronouncements alone that are subject to its rigid specifications. As you will see, it is this innovation that makes possible both the third and the fourth innovations.

The third innovation is a special kind of objectivity that can be imposed on scientific debate without being upended by the essential subjectivity of scientific reasoning—a kind of objectivity that is consistent with everything I have said in previous chapters about scientists' moral frailty and the multifariousness of their plausibility rankings.

The iron rule's negative clause is its fourth innovation. The natural philosophers cared a great deal about their theories' power to explain natural phenomena. They also cared about their theories' philosophical integrity, theological purity, and formal beauty, and they were ready and

The Four Innovations That Made Modern Science

1. A notion of explanatory power on which all scientists agree
2. A distinction between public scientific argument and private scientific reasoning
3. A requirement of objectivity in scientific argument (as opposed to reasoning)
4. A requirement that scientific argument appeal only to the outcomes of empirical tests (and not to philosophical coherence, theoretical beauty, and so on)

willing to make a case for their views from every one of these perspectives. The iron rule, however, permits nothing but matters of explanatory power, nothing but a theory's ability to account for the observable, to determine the course of scientific argument. Theology, philosophy, even beauty are strictly off limits. Scientists, if they choose to dispute, are obliged to do so in the empirical manner, by running tests in accordance with the rule's prescriptions.

In this chapter and the last, you have seen how modern science's procedural consensus, established by the iron rule, spurs the creation and compilation of the sort of rich, intricate, revealing data that has enabled scientific progress in a Tychonic world. It is now time to see how the four innovations invest the iron rule with its ability to create that consensus.

CHAPTER 6

Explanatory Ore

*How explanation became objective, yielding the material
to forge an iron rule that says the same thing to every
scientist (the iron rule's first innovation)*

THERE IS AN undiscovered island somewhere in the southern oceans, a scrap of land whose people have yet to make contact with the modern world. Call it Atlantis. If it does not exist, invent it: it is needed for philosophical purposes.

The Atlanteans, though few in number, have a scintillating civilization. Poets, architects, playwrights, and mathematicians mill about their court and their marketplaces, along with natural philosophers who aim to understand the workings of the visible world: the planets, light, falling objects, living things. But like the ancient Greeks—or the Ming dynasty Chinese or the Elizabethan English—the Atlanteans lack modern science.

After one navigational error too many, you find yourself washed up on the Atlantean shores. Your hosts are thrilled and perplexed by their mysterious visitor. You tell them about vaccines, antibiotics, and laparoscopic surgery. You show them your smartphone—a remarkable object, though the signal strength is questionable. You describe a world of effortlessly abundant food, half-mile-high towers, and flying machines.

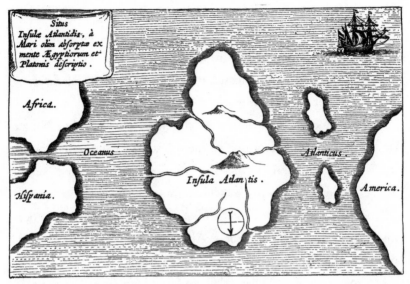

Figure 6.1. The location of Atlantis, as imagined by the scholar Athanasius Kircher (1602–1680). He places it a little too far to the north.

Better still, a world that understands the true nature of light, the principles that move the planets, and the code that programs life. The Atlanteans want it all. Can you perhaps teach them to do science?

Why not? It will help pass the time. The Atlanteans bring out their natural philosophers, eager pupils in the inaugural session of your first seminar: "Introduction to Modern Science."

Day one: the iron rule. There is no need to persuade the Atlantean philosophers that they should search for theories of great explanatory power; they've been doing exactly that for centuries. What's news is that exploiting this insight to make scientific progress will depend far more heavily on exacting measurement, on painstaking laboratory toil, than on deep thought. In our Tychonic world, it is by accounting for the fraction of an inch, the sliver of a degree, that the truth will make itself known.

Day two: you make an astounding discovery. The Atlanteans consider a theory to explain a fact whenever that fact rhymes with the theory. According to these instinctive poets of nature, the hypothesis that "things made of earthly matter seek to find the center of the universe" explains why "he who topples from a lofty crag comes out much the worse" but not why "straws tossed into a storm fly near and far away," since *worse* rhymes with *universe* but *away* does not.

What could the Atlanteans be thinking? Perhaps they imagine that the world is run according to the dictates of a "Great Poet in the Sky." This supernatural versifier has laid down certain edicts, such as "Things made of earthly matter seek to find the center of the universe," and ensures through direct intervention in the world that events occur only if they rhyme with the edicts. Although this is a bizarre view, it is not incomprehensible. "I see," you say to the Atlanteans. "Like us northerners, you Atlanteans believe that explaining an event is a matter of finding out what made it happen. But unlike us, you believe in gods who make things happen by way of rhyming rather than gravity."

No, that's not it at all, say the Atlantean philosophers. There is no celestial sonneteer. Indeed, there is nothing that "makes things happen." The world simply is. But its state of being can nevertheless be understood, its workings can be explained, by grasping how it all hangs together—as a poem. For the Atlanteans, understanding the universe is a matter of literary interpretation.

Bringing science to Atlantis is going to be harder than you supposed. You clear your throat. There are a few things you need to know about explanation, you say. Or at least about the kind of explanation that supplies the raw material for the iron rule. Most important of all, explaining events is a matter of finding their causes, not of rhyming. You cannot discover true theories by scrutinizing word endings! The Atlantean philosophers mutter among themselves, casting occasional pitying glances in your direction: such splendid technology; so little understanding. One speaks up: "No, I'm afraid it is *your* theories that must be incorrect." Who are the teachers and who is the pupil?

If only you could open your eyes and . . . ah, the real world! Electric lighting, anesthesia, and broadband internet! It was all a dark fantasy. Wasn't it?

It is not so far from the truth, contend the many historians of science who look back in time and see an archipelago of ideas about understanding, a chain of cultural islands stretching over the centuries, each with a proprietary and parochial concept of what it is for a theory to provide an adequate explanation of patterns of everyday events and experimental outcomes. Each era has its own rules for making sense of the world; there are, writes the contemporary historian Peter Dear, "no timeless, ahistorical criteria for determining what will count as satisfactory to the understanding."

Cast your mind back to Thomas Kuhn, early in his career, attempting to comprehend Aristotle's physics. He encountered some peculiar ideas. Aristotle regarded an object's location in space as though it were a property like color; moving, then, is a bit like changing color. As a consequence, there cannot be an empty location in space: locations are properties of things, so where there is a location, there is a thing, in the same way that where there is a color there has to be a thing to bear that color. In fact, according to Aristotle, there is no such thing as space; there are only "places," and they are all filled. (He was perhaps the first theorist of inner-city parking.) Further, characteristic patterns of motion, such as heavy things' tendency to fall to earth, are given an explanation that seems biological or even psychological: the "natural end state" for the heavier varieties of inert matter is to occupy the exact midpoint of the universe. All things, animal, vegetable, and mineral, seek out their natural states. Thus, inert matter heads directly for the center of the world, which is also the center of the earth. None of this, at first, made any sense to Kuhn; Aristotle's claims seemed so obscure as to be unintelligible.

Then he looked out the window and everything, as he said, "fell into place." He achieved the ability to see the world as Aristotle did, to operate with Aristotelian principles for explaining the world—principles that were so strange as to be almost Atlantean. "Now I could understand why [Aristotle] had said what he'd said." That experience inspired Kuhn's conception of scientific progress as a series of leaps from framework to framework, with each such "paradigm shift" bringing a new explanatory system that mystifies scientists educated in an older way of looking at the world.

A paradigm is more than an explanatory framework; it is a complete recipe for doing science, including goals and methods as well as modes of understanding. Modern historians are skeptical, as you saw earlier, that scientists' behavior conforms to any schema as totalizing as a Kuhnian paradigm. But like Peter Dear, many hold on to the explanatory component of Kuhn's picture even as they discard the rest, endorsing along with Kuhn an idea that might be called *explanatory relativism*. The thesis of explanatory relativism says that each age or school has its own standards for explanation and understanding and that one age's standards typically make little or no sense to the thinkers of another age.

In this formulation, "make no sense to" does not mean merely "appear implausible to." It means that explanations from another time do not seem to be explanatory at all—just as we consider the Atlanteans' rhyming scheme to be, however beautiful, quite incapable of conveying true understanding of why things happen. Untimely explanations, then, are off the table altogether, disqualified from inquiry before the game begins. Only a skilled historical interpreter, as Kuhn took himself to be, can appreciate their virtues, and even then perhaps only by way of some unbidden, almost mystical epiphany.

If the thesis of explanatory relativism is correct, then there can be no consensus, over time, about different theories' comparative explanatory success, since "success" means something completely different for each theory. In that case there would be no way to establish the sort of procedural consensus that, I have argued, turns the crank of the knowledge machine. Explanatory relativism is therefore incompatible with the iron rule; were it to be true, modern science could not exist.

It once was true. But then, during the Scientific Revolution, it was deposed. This chapter will explain how relativism undermines consensus and how it came itself to be undermined, clearing the way for a new form of inquiry governed by the iron rule.

It is the fourth century BCE. You are wandering an island less fictional than Atlantis—ancient Greek Lesbos—in the company of Aristotle himself, observing the diverse aquatic life in its lagoon. (In the 340s BCE, the great philosopher spent a number of years on Lesbos and in the nearby mainland city of Assos, researching his treatises on biology.) There is a great deal on which the two of you might, by close and regular observation, come to agree: that some fish make grunting or piping noises, that male octopuses have a specialized tentacle for copulation, that dolphins do not lay eggs but rather give birth to live young, that the lagoon is rich in oysters, and that in winter many of its fish migrate into the warmer open sea. You might dissect a cuttlefish together, as we know Aristotle did, and find that the intestinal tract runs from the mouth to the stomach, then loops back to exit just below the head.

When you come to explain the cuttlefish's anatomy and behavior, however, the common ground beneath your feet starts to shift uncertainly. What does the cuttlefish heart do? You might think that it pumps blood, but Aristotle has other ideas: the heart is the primary source of blood, but also the seat of sensation—pleasure,

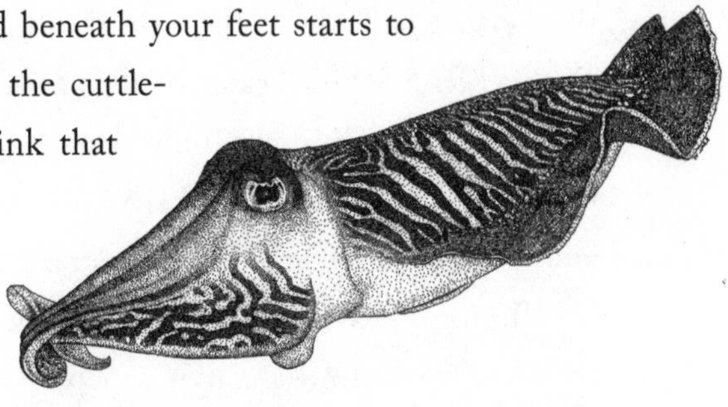

Figure 6.2. The common cuttlefish, *Sepia officinalis*.

pain, and sense experience—and in higher animals, the organ of thought. (On this matter Aristotle was at odds with many of his contemporaries, who located thought in the brain.) That is a disagreement, but no different from Capulet and Montague's disagreement about the nature of heat. Surely empirical testing will decide what's correct.

To help the process along, you have brought a surprise for your philosophical companion: a scalpel, dyes to accentuate cell walls and other biological boundaries, and a microscope. With the help of these newfangled instruments, you'll be able to understand the operation of a cuttlefish heart by examining the microscopic structure of its parts and figuring out how that structure, guided by the mechanical principles of physics, causes the behavior of the whole.

"Interesting," says Aristotle, stroking his well-tended beard. "But like the ideas of those old materialists Empedocles and Democritus"—here he refers to his predecessors and rivals, now long gone—"utterly implausible." He cocks a critical eyebrow. "Their picture of the world was so simple. And so simplistic: everything we see, they taught, is caused by the blind laws of physics pushing around dumb matter. If they were right, we would find monstrousness and malfunction everywhere. Perhaps just by chance an occasional collection of mechanical parts would work together like an organic body, feeding and breathing and turning out little duplicates of itself. But these would be by far the exceptions among the chaos." With a sudden, uncharacteristically loose and expansive gesture he takes in the hills, the water, the entire island. "Look at all this life! See how beautifully organized it is. Every plant, every animal, every branch, every limb, every organ, works carefully, regularly, precisely in its own way to thrive in its natural place in the world. It is anything but accidental. Physics alone cannot explain life."

The only way to explain the cuttlefish's characteristic form and activities, he continues, is to posit a property of the organism as a whole, its *psuche*, often translated as "soul." It is not what it sounds like: Aristotle's *psuche* is not an immaterial thing detachable from the animal in question. But nor is it a material part of the animal, like a leg, a brain, or a heart. It is rather something that subsists in the animal and causes it to have its overall form. The nature of the *psuche* is to organize things for the ultimate good of the animal. Thus, the animal's parts operate so as to perform functions that, working together, cause the animal to grow, flourish, and proliferate.

Every modern interpreter of the Aristotelian *psuche* finds in it a puzzle. It is, says Aristotle, the form of the animal in its body, which suggests that the *psuche* is simply the physical structure of the creature. That cannot be what Aristotle means, however, or else he is no better off than the materialists he criticizes, attributing the viability and efficiency of organisms to the fact that they happen by chance to have the ideal physical organization for the job. At the same time, it seems, the *psuche* is not something above and beyond the body. It is the physical form of the body, yet it also explains and sustains that form.

Reach for an understanding of the *psuche*, and it flits away into the depths. It will not be captured by a mind like yours. That is the characteristic symptom of explanatory relativism. The only intelligible explanation of life for Aristotle is to you unintelligible—barely better than the works of the explanatory poets of Atlantis, who sit in their libraries pondering rhymes for "cuttlefish."

So there stand two figures at odds on the shores of the lagoon: you with your microscope and Aristotle with his conception of natural order. You want to use your microscope to settle your differences. It will reveal a physical structure in cuttlefish tissue that is indiscernible to the naked eye, you say, yielding invaluable clues about the workings of the heart and other organs. "Futile," responds Aristotle. It is not that physical structure has no role to play in bodily functions (although Aristotle does in fact deny that organic tissue has microscopic structure). It is rather that physical structure is not at all the right kind of thing to account for biological function. Even if it were to explain how a cuttlefish moves, it could not explain why the movement is tuned so finely to promote life. Only *psuche* can do that.

In Aristotle's eyes, your program of inquiry—slicing cuttlefish parts into finer and finer pieces to hold up to the light—is not sophisticated science but plain butchery. Aristotle's program, meanwhile, looks not much better from the perspective of your own explanatory commitments. How can you detect a *psuche* and probe its properties and operations? Or more exactly—since it is easy to detect the physical form in which the *psuche* inheres—how can you detect and measure the explanatory aspect of the *psuche* that goes beyond mere physical form? To Aristotle, the answer is obvious: the appearances and activities of every organism are explained by its *psuche*. So evidence for the nature of the *psuche* is anywhere that a biologist might care to look. To you, with a different notion of biological explanation, that sounds not so much like empirical inquiry as an exercise in begging the question.

A disagreement about explanatory systems thereby leads to intellectual impasse. For Aristotle, the microscope is no more useful for biology than for astronomy: what it might detect, if anything, can be seen in advance to lack the power to explain life. For you, the explanatory power of the *psuche* is a mystery. You'd like to use your instruments to get an independent grip on the thing, but the only way to map its structure is through its explanatory ramifications, which are precisely what to you are so obscure.

By fomenting communication breakdowns of this sort, explanatory relativism presents an insuperable obstacle to the kind of procedural consensus essential for modern science, a consensus that assures scientists they are all playing the same game, freeing them from endless debates about the explanatory rules and so directing every last flicker of their investigative ardor toward observation and experimentation. The iron rule therefore outlaws relativism, cutting off all such disagreements by providing a fixed conception of explanatory power to which all scientists must subscribe—

modern science's first great innovation. Because scientists agree on what can explain what, they agree on how to go about testing any hypothesis, even a hypothesis couched in esoteric or controversial terms such as *psuche*, entropy, or the obscure terminology of quantum mechanics.

The explanatory accord was hard-won. To see how it came about, we will have to leave Aristotle on Lesbos and travel to northwestern Europe, two thousand years later. There we will entertain the ideas of two superlative thinkers. One was the last of the great natural philosophers, devising theories of matter and its behavior that conformed to an idio-syncratic, metaphysically regulated scheme of explanation yoked to the zeitgeist. The other destroyed these theories, and in so doing, shattered the dominion of philosophy over scientific explanation and overthrew explanatory relativism itself. The arc curving from the one intellectual life to the other is the story of the smelting of the new explanatory iron.

In 1618, as Francis Bacon worked on the manuscript of his *New Organon*, a young French student turned his back on law school and joined the Dutch army just as war erupted on the other side of Europe in Bohemia. Within a year he had left the Dutch and joined the allied forces of Maximilian of Bavaria, a decision that took him to the German states that made up the bulk of the Holy Roman Empire. There, most of the fighting and most of the dying in this devastating conflict would take place. By its end, the Thirty Years' War would have killed nearly half of the population in Bavaria, a third in Bohemia, and at least a fifth of the people of Central Europe as a whole—a greater percentage than any other European war in history to this day.

In the dull and peaceful months before the German princes were drawn into the struggle, the French soldier found himself at a loose end while billeted in the city of Ulm, on the Danube in what is now southern Germany. It was November; he was cold. He shut himself into a well-heated room and began to think; in the course of his cogitations, he later related, he "discovered the foundations of a marvelous science."

That night he dreamed three times. First came a dream of fear: specters, paralysis, a blasting wind in which he could not walk upright. Second came a dream of undirected power: fire and thunder. Third came a dream of books and learning, of the beginning of a journey, of the unity of all knowledge. In the following days, he resolved to devote his life to building his marvelous new science. René Descartes became a philosopher.

In short order, Descartes left the army and so left the war. He lived as a recluse in France and the Netherlands for the next three decades, while Europe chewed itself up around him. In his solitude, he wrote about mathematics, physics, philosophy, God, vision, thought, and, in the end, "the passions of the soul." He explained depth perception, the refraction of light, and the nature of the rainbow, and he pioneered the use of what we now call Cartesian coordinates in mathematics. ("Cartesian" means "proposed or invented by Descartes.") What was perhaps his grandest and most ambitious project, however, was also one of his first: a description of the structure of the universe, which he finished in 1633. Descartes called this book *The World*.

All the activity that makes up the world, proposed Descartes, from the simple orbits of the heavenly bodies to the intricate functioning of the human body, is the movements of chunks of matter, and these chunks interact in one way only: by direct contact, or in other words, by pushing each other around. The planets may look like they are out there hurtling through empty space. But it can't be so. If something moves, it is because something else forces it to move, and the sole way to transmit such a force is direct physical contact. The planets maintain their orbits around the sun, Descartes hypothesized, because what appears to us to be empty space is full of imperceptibly tiny particles that are themselves revolving around the sun, carrying the planets with them.

The universe, Descartes ventured, is packed with giant globules of matter rotating in this way; at the center of each is a sun around which its own proprietary planets revolve. To us, these suns are the distant stars (Figure 6.3).

Gravity on earth, since it is a matter of force, must also be caused by something in direct contact with the bodies that experience gravitation.

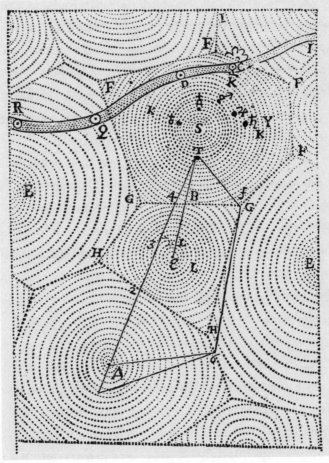

Figure 6.3. Descartes's universe. Each of the polygonal cells is a great rotating globe of matter with a star at its center. Our sun is marked S; the planets are labeled with their astrological signs (♂ for Mars, ♃ for Jupiter, and so on). A comet (☉) blazes an irregular path through the "north" of the solar system.

Descartes dreamed up the following intricate story. All bodies, he reasoned, have a tendency to "fly away" from the earth by a kind of centrifugal force. Because air particles are more "agitated" than the particles of which heavy objects, such as rocks and humans, are composed, their centrifugal tendency is more powerful: your body is trying to fly away into the sky, but the air around you is trying even harder. Consequently, if you are not on solid ground, the air below rushes up past you, pushing you down into the space that it previously occupied; you fall, then, because the air is pushing you downward, and this downward pressure is greater than your body's tendency to move upward. The same mechanism keeps you from rising in the first place. It is as though you are in a crowd shoving your way toward the exit to escape a fire, but because the other members of the crowd are stronger and more determined than you, they elbow their way past, keeping you firmly in place—clamped to the ground.

The Cartesian explanation of gravity is ingenious but baroque, and indeed, rather awkward. If the awkwardness embarrassed Descartes, he didn't show it. Perhaps that was because he regarded his collision-driven physics as far more than a plausible empirical proposal. It had to be true, he believed, because he could prove that the world couldn't possibly work in any other way. His demonstration was not scientific in our modern sense, but rather philosophical and theological. It began with a metaphysical argument.

Space, he claimed, is by its very nature extended, that is, spread out. Empty space would therefore be an extended nothing, but "nothingness cannot possess any extension." It follows that there can be no empty space; space is necessarily filled with matter (as Aristotle, too, maintained). Indeed, in some sense, according to Descartes, matter and space are one and the same thing, described in two different ways. "It is no less impossible that there be a space that is empty," he wrote, "than that there be a mountain without a valley."

If space is completely packed, you might think, nothing could ever

get started. But when God created the universe, said Descartes, turning from philosophy to theology, he set all its matter in motion, creating
the great spinning globules pictured in Figure 6.3. From that point on,
things followed the principles of Cartesian physics. The matter in a glob-
ule continued to rotate around its central point unless other matter got in
the way—in which case there was an impact having the power to send
the colliding matter off in new directions. (Because all space is filled with
matter, Descartes's collisions are more like jostles and shoves than projec-
tile impacts. I think of certain New York City subway lines at rush hour.)

Every departure from a steady state of motion, then—every meaningful change—is caused by direct contact. This is true, needless to say,
of manifest collisions, such as a cannonball's crashing into a fortress wall.
But with God and philosophy on his side, Descartes sought to show that
change works that way for everything else as well. Gravitational attraction, for example, must somehow be a matter of our being pressed directly
to the earth, even if there is no tangible sign of anything doing the pressing. Descartes therefore concocted the story in which gravity is effected
by small particles pushing aside larger ones in their skyward stampede.

Likewise, visual perception must be powered by collisions. Light
strikes the retina, Descartes proposed, which then signals the brain by way
of some intricate hydraulic machinery that pumps "animal spirits" around
the body. Collision underlies all other bodily functions for the same reason. Of necessity, every biological process is mechanical, and every mechanism is implemented by flows of matter interacting by way of impact.

The Cartesian and the Aristotelian views could hardly be more
different—in their details but even more so in their fundamental explanatory principles. According to Aristotle, as we saw, each kind of matter
has a natural motion. The natural motion of heavy matter, of things made
of the elements earth and water, is toward the center of the universe, that
is, toward the center of the globe on which we live. The natural motion
of the stars and planets, which are made of a fifth element—closer to

divinity than air, fire, earth, and water—is the circle. Thus, metal falls to the ground while the moon eternally orbits the earth; in both cases, they do what it is in their nature to do, unaided.

Descartes explains these same motions in terms of pushes and shoves. Even the planets, set on their way by God, require the pressure of other, unseen particles to keep them on their circular paths. To posit any other cause of motion, Descartes thinks, falls short of true science. Close examination of such a hypothesis, he would say, will reveal links in its proposed explanation that are no better than rhymes—pleasing to the mind, perhaps, but failing to establish a genuine explanatory connection between cause and effect.

The same goes for plant and animal biology. Descartes conceives of organic bodies as machines, while for Aristotle, as we saw, organisms are distinguished from machines by their possession of a *psuche*, an explanatory principle that subsists in the body's physical structure and explains the regularity and purposiveness of the actions brought about within and by that structure.

Descartes dismisses Aristotle's explanations as unintelligible; Aristotle would dismiss Descartes's as hopeless. To each the other's proprietary explanatory framework looks confused, empty, incomprehensible, absurd—as intellectually alien as the Atlantean rhyming scheme.

Yet at the most abstract level, Descartes and Aristotle have something of great importance in common. Both call upon higher powers—most notably, the power of philosophical reason—to determine what can legitimately explain the movements of matter and living things. It is this philosophizing that digs them into diametrically opposed positions. Their explanatory principles settled, their theories of physics and biology can only remain perpetually at odds.

What next? If Aristotle had surveyed the intellectual battlefields of the 1600s, he would surely have exclaimed, "Not again!" The atomists and other materialists, vanquished in fourth-century Athens, had returned and must be fought once more. It was just as Bacon said: the same arguments, going round in circles forever.

And so it might have continued, had not something new appeared to break the cycle and to direct the investigation of nature onto a wholly new trajectory, along which profound disagreements about explanatory standards would become quite unknown.

DESCARTES DIED IN 1650. The war in Europe was over. Germany lay in ruins; France was triumphant; England, having executed its king, Charles I, was a republic. Over the silent battlefields settled a peace born of exhaustion. In the universities and learned societies, by contrast, the gravity wars were poised to begin.

Isaac Newton was 7 years old. By the age of 11, he was ranked the second worst pupil in his class at King's School in Grantham, Lincolnshire. At some point, however—possibly in reaction to his bullying by the third worst pupil—he threw himself into his studies, which drew him to Cambridge University's Trinity College. Like Descartes, he preferred his own company, the better to work without interruption: "Truth is the offspring of silence and unbroken meditation." Unlike Descartes, who died in Sweden, he never left the country of his birth.

Indeed, Newton barely left the university, and he was still at Cambridge when, 25 years later and now the Lucasian Professor of Mathematics, he published his greatest work: the *Mathematical Principles of Natural Philosophy*, usually called (after its original Latin name) the *Principia*. The world had seen nothing like Newton's system before. It had the scope of the most ambitious physics, such as that of Aristotle and Descartes, the mathematical exactitude of the best astronomy, such as that of Ptolemy and Kepler, yet its core could be written down in a handful of simple formulas: Newton's three laws of motion and his law of universal gravitation.

Whereas Descartes held that all physical causation is by direct contact, Newton appealed to a "force of gravity" that pulls one object toward another massive object, apparently without any intervening mechanism. Many of his contemporaries saw in Newton's new force a gaping theoretical hole. How could gravitation be taken seriously unless there was a physical means, such as currents of interstellar fluid or collisions with tiny particles, to make it happen? Causal forces between planets made no sense without a tangible thing or stuff connecting them, communicating a pushing or a pulling or some other kind of visceral "oomph." The notion that causation could happen otherwise—that one planet could simply act on another at a distance, as though extending a ghostly grip of arbitrarily long range—seemed preposterous. The mathematical and metaphysical prodigy G. W. Leibniz, writing around the time that Newton issued the second edition of his magnum opus in 1713, dismissed Newton's theory as "barbaric physics," testimony to the sad fact that "people love to be returned to darkness."

In the decades after Newton, however, the idea of "action at a distance," of celestial bodies influencing one another's movements across empty space with no connecting causal links, became familiar and acceptable. If anything, it was causation by collision that seemed suspicious. Why should two objects, coming into contact, refuse to interpenetrate? Some sort of repulsive force, operating at a very short range, must prevent them from overlapping. If so, what appear to be genuine impacts are in fact merely unconsummated flirtations with physical contact: particles do not literally touch but at most come extremely close before irresistible countervailing forces fling them apart. When the great German philosopher Immanuel Kant wrote about physics a century later, in 1786, he claimed that it was incomprehensible that objects could act on each other except by way of brute forces, projected across space in accordance with something like Newton's law of gravity.

As I have related the story so far, this sounds like yet another chapter in the chronicle of explanatory relativism, yet another example of intellectual history as a series of discontinuous jumps from one explanatory framework to another. In the seventeenth century, causation had to be communicated by contact; anything else was darkness. Around 1700, according to the relativist story, Newton successfully pushed Descartes's conception of legitimate explanation aside, substituting his own—just as Descartes had earlier routed the Aristotelian approach to explanation. From then on, contact looked philosophically incoherent; action at a distance became the only logically acceptable way to make things happen.

That's how Kuhn spins the history of gravity in *The Structure of Scientific Revolutions*. But his relativist interpretation is wrong. Newton did not change his age's conception of explanation but rather something deeper still—he changed its conception of empirical inquiry.

To understand how, let us take a closer look at Newton's own interpretation of his method, laid out in a postscript to the *Principia*'s second edition of 1713. There Newton summarizes the fundamental properties of gravitational attraction—that it increases "in proportion to the quantity of solid matter" and decreases in proportion to distance squared—and then continues:

> I have not as yet been able to deduce from phenomena the reason for these properties of gravity, and I do not feign hypotheses. For whatever is not deduced from the phenomena must be called a hypothesis; and hypotheses, whether metaphysical or physical, or based on occult qualities, or mechanical, have no place in experimental philosophy. . . . It is enough that gravity really exists and acts according to the laws that we have set forth and is sufficient to explain all the motions of the heavenly bodies and of our sea.

In this famous passage, the essential aspects of the iron rule make a decisive appearance in world history.

Look at what Newton says about the mechanism for gravitation. You can see at once that he was not an ardent promoter of action at a distance. He allows the possibility that gravity is effected mechanically—by contact or collision—and also the possibility that it is effected by "occult qualities," that is, by some form of nonmechanical causation, such as action at a distance, the kind of thing that philosophers such as Leibniz found to be no more intellectually respectable than witchcraft. The mechanical and occult possibilities are, in this passage, put entirely on a par. Both are mere "hypotheses"; we presently have no way to tell which, if either, is true.

It does not matter. What does matter is that gravity "is sufficient to explain," with great quantitative accuracy, the observed phenomena: the orbits of planets, the paths of comets, the cycle of tides, and the arcs of deadly projectiles. To a philosopher such as Descartes, this modest proposal would make no sense. Gravity cannot explain anything unless it passes the test for being a genuine explainer—the collision test. To put the Cartesian view another way, to explain something means grasping its cause, and to grasp its cause means to comprehend the sequence of pushes or collisions that bring it about. Without that comprehension there is no understanding, no real explanation.

Newton, by contrast, pioneers in this passage a philosophically far shallower notion of explanation, on which a phenomenon is explained simply by deriving it from the causal principles of gravitational theory—that is, from the mathematical principles laid out in the *Principia*. Shallow explanation does not require the explainer to grasp the implementation of the principles. More important still, it does not require the principles to pass any philosophical test or to conform to the explanatory prescriptions of a Kuhnian paradigm. Gravity might turn out to be transmitted by impact, but equally it might turn out to be authentic action at a distance. Either way, Newton maintains, we have a "sufficient explanation" for the purposes of empirical inquiry, or as Newton calls it,

"experimental philosophy." Newtonian explainers, like Popperian falsi-
fiers, prove their worth by conforming closely not to intellectual precepts
but to observed phenomena. The shallow conception of explanation
thereby frees scientific theory builders to try just about anything, however ideologically abhorrent, in their attempts to explain the phenomena.

With this act of liberation, Newton escaped the endless circles of
explanatory relativism and gave scientists a "timeless, ahistorical criterion" for explanatory power to serve as the raw material for an iron
rule that dictates, in turn, a fixed criterion determining what counts as
a legitimate empirical test. A scientific theory postulates some causal
principles; what it explains is whatever can be logically derived from
those principles. That is a standard for explanation that means the same
thing in all places and all times.

To appreciate fully what is objective and what is subjective in scientific practice, however, we need to distinguish more carefully than Newton did between theories and theoretical cohorts. Theories, as you now
know, explain very little on their own; very little can be derived from
a theory without the assistance of auxiliary assumptions. Einstein's or
Newton's theories of gravity, for example, did not on their own predict
the positions of the images of stars on Eddington's photographic plates.
For that, Eddington needed to add information about the positions of
the telescopes, the times that the photographs were taken, the positions
of the stars themselves, the assumption that the telescopes were functioning correctly, and certain theoretical posits about light. When surrounded by statements of background conditions and other hypotheses
such as these, a theory forms a theoretical cohort and generates specific
predictions and explanations.

The iron rule's Newtonian criterion for shallow explanatory power
applies to cohorts rather than to bare, unaccompanied theories. It provides an objective and ideology-free test for whether a cohort explains a
phenomenon: Can the phenomenon be derived from the cohort's causal

facts and principles? Aristotle's centralizing urge; Descartes's democratic jostling; Newton's matter calling to matter across empty space—each qualifies as a causal principle that might make sense of gravitation, and each is warmly welcomed into science. All the iron rule asks is that such principles and their cohorts entail what is actually seen to occur and nothing that is seen not to occur. The modern scientific standard for explanation is as empirically demanding as it is philosophically lax.

Is the ineffable Aristotelian *psuche* scientifically admissible as an explanatory construct? Provided that it can be supplied with auxiliary assumptions to form a cohort that has definite observable consequences, yes. Even if *psuche* is in some sense unintelligible to the twenty-first-century mind, from the iron rule's point of view it is a potential scientific explainer all the same, just as long as its causal implications are, in a suitable theoretical context, explicit and sufficiently detailed.

Though permissive, the iron rule's conception of explanation is nevertheless not a matter of "anything goes." The Atlantean view of explanatory power does not qualify. Rhymes won't get you scientific glory; your theory has to tell a causal story.

By the time Newton died in 1727, it was already clear that he was, even among the great luminaries of the era, something special. At his funeral the pallbearers were dukes, earls, and the Lord Chancellor of England; the young French Enlightenment thinker Voltaire, who was present, reported that he was "buried like a king." Tributes to his brilliance lit up the eighteenth century with verbal fireworks. The Scottish philosopher David Hume described him as "the greatest and rarest genius that ever wrote for the ornament and instruction of the species"; Voltaire held him to be "the greatest genius that ever existed." The Marquis de l'Hôpital, a French mathematician, supposedly went further still: Newton was "a celestial intelligence entirely disengaged from matter."

Such intimations of near-divinity had begun with a poem that Edmond Halley wrote for the first edition of the *Principia*, ending with the line, "No closer to the gods can any mortal rise." The English artist George Bickham put the words into pictorial form in a 1732 engraving showing Newton as the sun surrounded by angels, muses, and putti. Around the same time, the Venetian painter Giovanni Battista Pittoni

Figure 6.4. Newton at home with the gods. On the left, *Sir Isaac Newton*, by George Bickham the elder (1732); on the right, Giovanni Battista Pittoni's *Homage to Newton* (1727–1729).

conceived his *Homage to Newton*, in which an angel and the goddess of wisdom Minerva lead a procession of muses to Newton's shrine—a rococo confection featuring a colossal urn containing Newton's ashes and a reconstruction and commemoration of the optical experiment in which he used a prism to split light into its component colors (Figure 6.4).

What mattered above all for the spread of the iron rule's shallow conception of explanation, however, was not its advocate Newton's position as foremost among scientists, not even his metaphorical ascent into heaven, but, as the historian of science Mordechai Feingold puts it, his metamorphosis "into science personified." As a consequence, Feingold continues, "Newtonian science . . . became the model to emulate, the manifestation of 'superior knowledge' that summoned all other learning to reorient itself along similar lines." To be scientific simply was to be Newtonian.

So the investigation of nature changed forever. No longer were deep philosophical insights of the sort that founded Descartes's system considered to be the keys to the kingdom of knowledge. Put foundational matters aside, Newton's example seemed to urge, and devote your days instead to the construction of causal principles that, in their forecasts, follow precisely the contours of the observable world. The thinkers around and after Newton got the message, one by one.

Three centuries later, the iron rule's shallow conception of explanation continues to push science forward. There is perhaps no better illustration of its curious effectuality than the story of how the rule confronted the most perplexing scientific theory in history—quantum mechanics—and swallowed it whole.

WE COME TO LIFE suspended in fluid; we are then born into a world of solid things. These two ways that a substance can be, solid and fluid, have directed the scientific imagination from the earliest days. When Thales, around 580 BCE, hypothesized that the world was made of water, he opted for fundamental fluidity, as did his pupil Anaximander, who suggested air, which is also a kind of flowing stuff. The ancient Greek atomists, by contrast, thought that solidity lay at the bottom of it all, that every kind of matter, even water and air, is made of tiny, rigid, indivisible particles. The two views can be combined, as in the four-element theory of Empedocles the Sicilian: everything, he said, was composed of some mix of air, fire, earth, and water. Descartes, much later, found a different way of mixing solid and fluid: matter is rigid by its very nature but comes in many sizes, the larger chunks making up solid matter like the planets, while, at the other end of the scale, particles that can be infinitesimally tiny constitute a kind of dust that, like a fluid, flows into every available nook and crevice.

In the nineteenth century, more exotic forms of fluid and solid behavior were called upon to do explanatory service. As dramatized in my tale of Montague and Capulet, heat began the century as a fluid but ended it as the motion of solids—the disordered vibration of countless tiny particles. The same intellectual current revealed that gases, the most rarefied of tangible fluids, were themselves composed of microscopic solid particles traveling at high speed. Light, by contrast, was by 1900 thought to be a wave traveling through a fluid more subtle and ineffable than any gas, the electromagnetic ether. Inspired by this idea, a small but influential group of scientists proposed to analyze all solid matter as "ethereal." In the words of the German physicist Gustav Mie, writing in 1911:

> Elementary material particles . . . are simply singular places in the ether at which lines of electric stress of the ether converge; briefly, they are "knots" of the electric field in the ether.

After 2,500 years, it might have seemed that physics was sailing back to Miletus, where the ancient Greek philosopher Thales first made the bold surmise that all was liquid and liquidity was all.

The ship never arrived. By the 1930s, the great conceptual and metaphysical cornerstone that was the solid/fluid duality had been obliterated. Matter—now conceived as a mix of particles such as electrons and protons—was still in the picture, but thanks to quantum mechanics, it no longer behaved at all like the ordinary physical stuff of everyday experience, fluid or solid. It spent its days rather in a mysterious mode of being called "superposition."

Superposition was nigh unintelligible. It didn't matter—not to working scientists. What they cared about was shallow explanation, and quantum mechanics succeeded in that endeavor like nothing that had come before. The case of quantum mechanics, then, is the perfect exemplification of the supremacy, in modern science, of Newton's precept that, for the purposes of "experimental philosophy," derivation of the observable is the only kind of explanation that counts.

The first hints of superposition came in the initial decade of the twentieth century in a series of theories designed to make sense of puzzling phenomena involving radiation and light. Perhaps the most iconic of these was Albert Einstein's explanation of the photoelectric effect, in which light striking certain substances causes the emission of electrons. Light, Einstein posited, which usually behaves like a wave, in this case behaves like a stream of particles—"photons." Descartes had argued that light is made of particles. Nineteenth-century scientists had shown that it must be a wave. Now it turned out that it was not one or the other, but somehow both at once.

When the first full versions of quantum mechanics were formulated in the 1920s, it became clear that it was not only light that lived a double life: streams of electrons, universally assumed to be ordinary particles, were quite capable of behaving like waves. In fact, all matter appeared to be some sort of wave-particle chimera, as though ancient mythology had all along been truer to the hidden structure of the universe than ancient physics—as though the centaur, half horse and half man, or the sphinx, half woman and half lion, were better models of reality than Aristotelian substance or the atom.

It was a philosophical emergency. The nature of reality was at stake. What was quantum physics trying to tell us about the world? The question was intensely discussed in letters and in lab corridors, most of all those of Niels Bohr's Institute for Theoretical Physics in Copenhagen. Bohr was an enthusiast for quantum mechanics, which he had helped to pioneer; Einstein was a skeptic in spite of his own early contributions. The two scientists famously debated the adequacy of the theory at the fifth Solvay Conference on Physics and Chemistry in Brussels in 1927 and then again at the sixth in 1930. The physicist Paul Ehrenfest gave a sense of the spectacle:

> Bohr from out of philosophical smoke clouds constantly searching
> for the tools to crush one example after the other. Einstein like a jack
> in the box; jumping out fresh every morning. Oh, that was priceless.

Einstein had earlier written to another founder of quantum theory, Max Born: "Quantum mechanics is certainly imposing. But an inner

Figure 6.5. Participants in the fifth Solvay Conference, 1927, in Brussels. Bohr is in the second row on the very right; Einstein is at the center of the first row.

voice tells me that it is not yet the real thing." Ninety years later, physicists still have no alternative. It is beginning to look like quantum mechanics is indeed the real thing. The debates about its meaning continue. Yet the science rolls imperturbably on, as quantum theory has been extended since 1930 to apply to electromagnetic force, the interior of the atomic nucleus, and (though an undertaking still in its infancy) gravitation itself.

That is a remarkable scientific achievement. It is also a remarkable social achievement. In the world of ancient Greece or the pre-Newtonian seventeenth century, competing philosophies meant competing sciences. But quantum mechanics did not split into philosophical schools. Rather, even as Bohr, Einstein, and many other key figures at the Solvay Conference in 1927 philosophized furiously, the theory remained a unified set of ideas that "developed rapidly, disseminated very quickly, and met almost no resistance." The new textbooks written to explain quantum mechanics to students barely mentioned the disputes at all.

The interpretation of quantum mechanics was, in short, considered to be so much philosophical superstructure, perched on top of but hardly integral to the science below. The same is true today. Yet it is not as if the strangeness of the theory is invisible to scientists. Murray Gell-Mann, the discoverer of quarks, called quantum mechanics a "mysterious, confusing discipline." According to Roger Penrose, one of the late twentieth century's foremost mathematical physicists, quantum mechanics "makes absolutely no sense." "I think I can safely say that nobody understands quantum mechanics," remarked Richard Feynman. How can a theory be widely regarded both as incomprehensible and also as the best explanation we have of the physical world we live in?

The answer lies in the shallow conception of causal explanation, in which it is derivation rather than comprehension that is paramount. To inspect that answer more closely, let me take you wading in the shallows of quantum explanation. Though entirely nontechnical, the overview that follows will take a few pages. But I promise you that it's worth your time. What we'll see is a contrast between the obscure *nature* of super-position and its clear *consequences*. The natural philosophers, such as Aristotle and Descartes, cared about the consequences of their theories' causal principles, of course, but they also cared a great deal about their nature—about what the principles were saying about the metaphysical foundation of the world's causal structure. Modern science, by contrast, is oblivious to the principles' nature. Let it be as opaque as action at a distance was to Newton's adversary Leibniz, as Aristotle's *psuche* is to us: it is immaterial. All that concerns the iron rule is what kinds of things the causal principles are capable of bringing about. I will show you how quantum theory derives accurate predictions from a notion, superposition, that is quite beyond our human understanding.

Matter, says quantum mechanics, occupies the state called superposition when it is not being observed. An electron in superposition occupies no particular point in space. It is typically, rather, in a kind of "mix" of being in many places at once. The mix is not perfectly balanced: some places are far more heavily represented than others. So a particular electron's superposition might be almost all made up from positions near a certain atomic nucleus and just a little bit from positions elsewhere. That is the closest that quantum mechanics comes to saying that the electron is orbiting the nucleus.

As to the nature of this "mix"—it is a mystery. We give it a name: superposition. But we can't give it a philosophical explanation. What we can do is to represent any superposition with a mathematical formula, called a "wave function." An electron's wave function represents its physical state with the same exactitude that, in Newton's physics, its state would be represented by numbers specifying its precise position and velocity. You may have heard of quantum mechanics' "uncertainty principle," but forget about uncertainty here: the wave function is a complete description that captures every matter of fact about an electron's physical state without remainder.

So far, we have a mathematical representation of the state of any particular piece of matter, but we haven't said how that state changes in time. This is the job of Schrödinger's equation, which is the quantum equivalent of Newton's famous second law of motion $F = ma$, in that it spells out how forces of any sort—gravitational, electrical, and so on—will affect a quantum particle. According to Schrödinger's equation, the wave function will behave in what physicists immediately recognize as a "wavelike" way. That is why, according to quantum mechanics, even particles such as electrons conduct themselves as though they are waves.

In the early days of quantum mechanics, Erwin Schrödinger, the Austrian physicist who formulated the equation in 1926, and Louis de Broglie, a French physicist—both eventual Nobel Prize winners—wondered whether the waves described by quantum mechanics might be literal waves traveling through a sea of "quantum ether" that pervades our universe. They attempted to understand quantum mechanics, then, using the old model of the fluid. This turned out to be impossible for a startling reason: it is often necessary to assign a wave function not to a single particle, like an electron, but to a whole system of particles. Such a wave function is defined in a space that has three dimensions for every particle in the system: for a 2-particle system, then, it has 6 dimensions; for a 10-particle system, 30 dimensions. Were the wave to be a real entity made of vibrations in the ether, it would therefore have to be flowing around a space of 6, or 30, or even more dimensions. But our universe rather stingily supplies only three dimensions for things to happen in. In quantum mechanics, as Schrödinger and de Broglie soon realized, the notion of substance as fluid fails completely.

There is a further component to quantum mechanics. It is called Born's rule, and it says what happens when a particle's position or other state is measured. Suppose that an electron is in a superposition, a mix of being "everywhere and nowhere." You use the appropriate instruments to take a look at it; what do you see? Eerily, you see it occupying a definite position. Born's rule says that the position is a matter of chance: the probability that a particle appears in a certain place is proportional to the degree to which that place is represented in the mix.

It is as though the superposition is an extremely complex cocktail, a combination of various amounts of infinitely many ingredients, each representing the electron's being in a particular place. Taste the cocktail, and instead of an infinitely complex flavor you will—according to Born's rule—taste only a single ingredient. The chance of tasting that ingredient is proportional to the amount of the ingredient contained in the mixture that makes up the superposition. If an electron's state is mostly a blend of positions near a certain atomic nucleus, for example, then when you observe it, it will most likely pop up near the nucleus.

One more thing: an observed particle's apparently definite position is not merely a fleeting glimpse of something more complex. Once you see the particle in a certain position, it goes on to act as though it really is in that position (until something happens to change its state). In mixological terms, once you have sampled your cocktail, every subsequent sip will taste the same, as though the entire cocktail has transformed into a simple solution of this single ingredient. It is this strange disposition for matter, when observed, to snap into a determinate place that accounts for its "particle-like" behavior.

To sum up, quantum mechanical matter—the matter from which we're all made—spends almost all its time in a superposition. As long as it's not observed, the superposition, and so the matter, behaves like an old-fashioned wave, an exemplar of liquidity (albeit in indefinitely many dimensions). If it is observed, the matter jumps randomly out of its superposition and into a definite position like an old-fashioned particle, the epitome of solidity.

Nobody can explain what kind of substance this quantum mechanical matter is, such that it behaves in so uncanny a way. It seems that it can be neither solid nor fluid—yet these exhaust the possibilities that our human minds can grasp. Quantum mechanics does not, then, provide the kind of deep understanding of the way the world works that was sought by philosophers from Aristotle to Descartes. What it does supply is a precise mathematical apparatus for deriving effects from their causes. Take the initial state of a physical system, represented by a wave function; apply Schrödinger's equation and if appropriate Born's rule, and the theory tells you how the system will behave (with, if Born's rule is invoked, a probabilistic twist). In this way, quantum theory explains why electrons sometimes behave as waves, why photons (the stuff of light) sometimes behave as particles, and why atoms have the structure that they do and interact in the way they do.

Thus, quantum mechanics may not offer deep understanding, but it can still account for observable phenomena by way of shallow causal explanation, the kind of explanation, favored by Newton, that the iron rule cares about and uses to supervise modern science's procedural consensus. The theory was accepted so readily in the 1920s and 1930s because, for all the philosophical arguments surrounding its interpretation, it provided a clear and well-defined system for providing shallow explanations that had no serious rivals. Had Newton rather than Bohr debated Einstein at the Solvay conferences, he would perhaps have proclaimed:

> I have not as yet been able to deduce from phenomena the nature of quantum superposition, and I do not feign hypotheses. It is enough that superposition really exists and acts according to the laws that we have set forth and is sufficient to explain all the motions of the microscopic bodies of which matter is made.

Ehrenfest would have enjoyed that.

EXPLANATORY STANDARDS CHANGE with the prevailing philosophical outlook of the age; this thesis of explanatory relativism is inherent in Kuhn's vision of scientific inquiry as a parade of paradigms, each with a different way of making sense of the world. The thesis holds a grain of temporary truth. For millennia, explanatory standards did change: as the metaphysical wheel turned, so the explanatory cogs turned with it, from Aristotelian teleology to Cartesian collision.

Late in the seventeenth century, the philosophically unified machinery of inquiry in which the wheels and cogs rotate as one was spiked. Newton was the saboteur, and thus he was the chief architect of modern science's first great innovation. Rather than deep philosophical understanding, Newton pursued shallow explanatory power, that is, the ability to derive correct descriptions of phenomena from a theory's causal principles, regardless of their ultimate nature and indeed regardless of their very intelligibility. In so doing, he was able to build a gravitational theory of immense capability, setting an example that his successors were eager to follow.

Predictive power thereby came to override metaphysical insight. Or as the historian of science John Heilbron, writing of the study of electricity after Newton, put it:

> When confronted with a choice between a qualitative model deemed intelligible and an exact description lacking clear physical foundations, the leading physicists of the Enlightenment preferred exactness.

So it continued to be, as the development and acceptance of quantum mechanics, as unerring as it is incomprehensible, goes to show. The criterion for explanatory success inherent in Newton's practice became fixed for all time, founding the procedural consensus that lies at the heart of modern science. From the raw material of shallow explanation, the iron rule was forged.

CHAPTER 7

The Drive for Objectivity

ൟ

How the iron rule enforces objectivity in scientific argument
while allowing pervasive subjectivity in scientific reasoning
(the iron rule's second and third innovations)

O VER THE WALLS of a moonlit Spanish cemetery, one night in
the summer of 1868, climbed two figures with larceny on their
minds. Where better to pick up a few surplus bones? Their owners had
no further use for them, and the gravediggers had moved them aside to
make room for new clients. The pair found what they were looking for
"tumbled in confusion and half buried in the grass." They made their
selection and then their getaway.

Had the grave robbers been caught in the act, the police in the small
Aragonese town of Ayerbe would have been rather surprised to find that
they had in their custody the town doctor, Justo Ramón Casasús, along
with his 16-year-old son. Dr. Casasús was desperate. His son, failing at
school and perpetually in trouble with the authorities, was set on living
the bohemian life of an artist. The doctor's outlandish idea was to use
the specimens from the cemetery both as artistic subjects and as objects
of instruction: his son would, sketching the bones, learn the rudiments
of osteology and develop a newfound enthusiasm for a medical career.

By the time the son, Santiago Ramón y Cajal, collected the Nobel

Prize in Medicine in 1906, it seemed safe to say that the father's strategy had been a considered success. Cajal received the prize for discovering that the brain is constituted of discrete cells—neurons—along which impulses travel from head to tail. This "neuron doctrine" was the beginning of modern neuroscience.

Like many Nobels, the 1906 award was shared. The other winner was the Italian Camillo Golgi, who had developed a staining technique that allowed the delicate structures of nerve tissue to be seen clearly under a microscope. Cajal had taken this technique, improved it, and then used his peerless powers of observation to discern the individual neurons and their interrelationships, producing in the course of his inquiries many striking illustrations of neuronal structure (Figure 7.1).

At the Nobel ceremony in Stockholm, Golgi was the first to give his acceptance speech. He rose to his feet and began to denounce Cajal. The

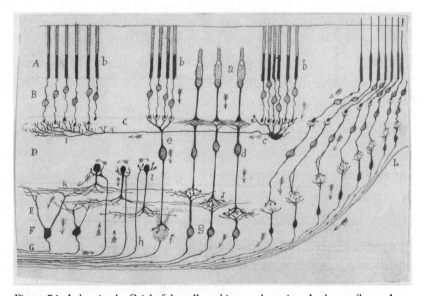

Figure 7.1. A drawing by Cajal of the cells making up the retina. At the top (layers A and B) are the light-sensitive cells; at the bottom (layer F) are the retina's output neurons. The signals travel in the direction of the arrows.

neuron doctrine, he argued, was wrong in every respect. Neurons were not in any sense isolated cells, but were joined to one another to form a great continuous network through which signals could flow in any direction. This was the "reticular theory" of the brain, an old idea that Cajal regarded as refuted by his meticulously made drawings—such as Figure 7.1, which shows a clear separation between neurons. But Golgi had his own drawings, which he displayed during his address, purporting to reveal the neural continuum that he described.

Cajal later repaid Golgi with withering scorn. He characterized the reticular theory as a lazy man's stratagem—"admirably convenient, since it did away with all need for the analytical effort involved in determining in each case the course through the gray matter followed by the nervous impulse." In his pursuit of the easy way out, Cajal further claimed, Golgi had presented images in support of the reticular theory that were "artificially distorted and falsified."

At bottom, Cajal accused Golgi of not looking properly at his own specimens, of allowing his preconceptions and psychic needs to interfere with his vision, of failing to present an objective account of what lay before him. Cajal considered himself, by contrast, to be a scrupulous scientific reporter. "Objectivity was at once the guiding and the unifying theme . . . for his career-spanning defense of the neuron doctrine," write the historians Lorraine Daston and Peter Galison. His depictions of the structure of the brain, Cajal thought, captured the true nature of the thing as surely as had his long-ago drawings of the stolen bones, of which he wrote, recording the success of his father's unorthodox program of persuasion, "Thirsting for the objective and the concrete, I seized eagerly the fragment of solid reality which [they] presented to me." That same sense of the objective and the concrete, he believed, had won him his Nobel Prize—and his contest with Golgi, as the neuron theory of the brain came to replace the reticular theory.

Cajal intended the barbs he fired at Golgi to sting, and they were able to do so because Golgi was no less committed than Cajal to objectivity in the presentation of scientific findings: he claimed that his own diagrams were "exactly prepared according to nature."

This clash of Nobels shows just how important objectivity is to scientists as a shared goal and hence as a communal standard against which a competitor's work can be measured and found deficient. It is not just Cajal and Golgi: everywhere you look, for a scientist to call a colleague's work objective is to bestow upon it high praise, while to deny its objectivity is to call into question its scientific worth.

But can this be anything more than talk if scientists' reasoning is essentially subjective? It can be far more than that. There is room for objectivity in science because it is limited to playing a rather peculiar and constrained—though crucial—role in regulating the procedural consensus imposed by the iron rule. The fashioning of that role constitutes what I have called modern science's second and third great innovations. To better understand scientific objectivity and its self-imposed limits, let's go back to a quintessential case of subjectivity, embarking one last time with Eddington on his expedition to test Einstein's theory of general relativity by observing the 1919 total eclipse of the sun.

THREE TELESCOPES, you will recall, were trained on the eclipse. The instrument on the African island of Príncipe barely saw the sun through the clouds. The two telescopes in Brazil gave conflicting testimony: one, the 4-inch telescope, saw light bent roughly as Einstein predicted, but the other—the Brazilian astrographic telescope—saw the bending predicted by Newtonian physics. Eddington discounted the results from the Brazilian astrographic on the grounds that its photographs were out of focus, while lavishing great care on his own equally obscure Príncipe photographs, teasing a pro-Einstein result out of the murk. He then convinced various British scientific luminaries to endorse his conclusion that the observation of the eclipse was a spectacular confirmation of Einstein's new theory. The astrographic anomaly was largely forgotten.

Eddington's behavior was not fraudulent, but his reasoning was partial and self-regarding. Enchanted by the elegance of Einstein's theory and intent on closing the rift between British and German science prised open by the Great War, he hungered for an Einsteinian triumph. So he was quick to conclude that the Brazilian astrographic had malfunctioned hopelessly even as he treated the badly compromised photos he had himself taken in Príncipe as imperfect yet highly informative. In his own eyes, he was reasoning his way toward truth, but

Figure 7.2. Albert Einstein and Arthur Eddington enjoy a quiet moment together at the University of Cambridge Observatory in 1930.

the plausibility rankings that went into his reasoning—in particular, his estimate of the probability of a systematic Brazilian astrographic breakdown—were twisted by his hopes and expectations for what that truth might be.

All this is, of course, just to repeat and to underscore the lesson learned from the collapse of the Great Method Debate: scientific reasoning is inflected always and everywhere with subjectivity. There is no place, apparently, for what the iron rule demands, an argument that is purely objective. Or is there?

Eddington and his collaborators reported the outcome of their expedition in a scientific paper with the informative if unwieldy title "A Determination of the Deflection of Light by the Sun's Gravitational Field, from Observations made at the Total Eclipse of May 29, 1919," which was published in a scientific journal, an official repository of research, called the *Philosophical Transactions of the Royal Society* (Figure 7.3).

Leaf through the pages of the eclipse paper, and you will find in the serried ranks of figures and careful calculations plenty that looks purely objective. In Figure 7.4, for example, the names and positions of the brighter stars surrounding the eclipsed sun are listed, along with Einstein's predictions for the apparent shift in the stars' positions. Even Eddington's fiercest critics will grant the accuracy and the impartiality of the information thereby presented. Those were the stars; those were their positions; and the Einsteinian predictions were calculated without error.

Or turn back to the second of the two tables that I extracted from the eclipse paper in Figure 2.3. There you have a faithful record of the 18 photographs taken by the Brazilian astrographic telescope, along with a calculation of the gravitational bending implied by each photo. The number at the bottom right-hand corner summarizes the results: the photographs seem to show that the "bending number" is 0.86, which is

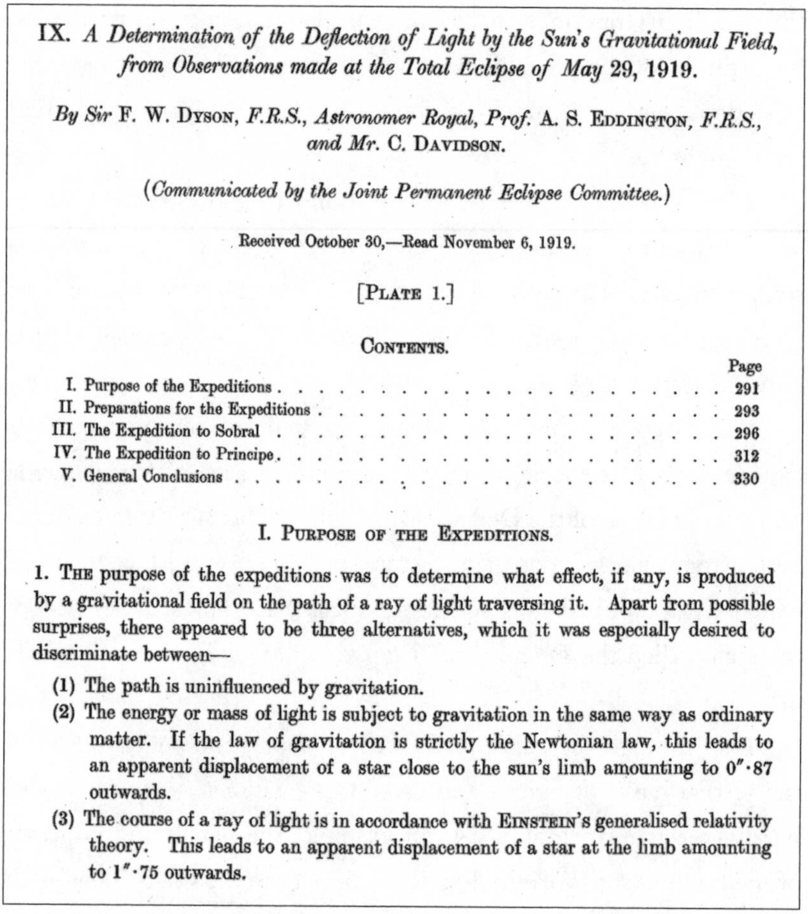

IX. *A Determination of the Deflection of Light by the Sun's Gravitational Field, from Observations made at the Total Eclipse of May 29, 1919.*

By Sir F. W. DYSON, *F.R.S., Astronomer Royal, Prof.* A. S. EDDINGTON, *F.R.S., and Mr.* C. DAVIDSON.

(*Communicated by the Joint Permanent Eclipse Committee.*)

Received October 30,—Read November 6, 1919.

[PLATE 1.]

CONTENTS.

I. PURPOSE OF THE EXPEDITIONS.

1. THE purpose of the expeditions was to determine what effect, if any, is produced by a gravitational field on the path of a ray of light traversing it. Apart from possible surprises, there appeared to be three alternatives, which it was especially desired to discriminate between—

(1) The path is uninfluenced by gravitation.

(2) The energy or mass of light is subject to gravitation in the same way as ordinary matter. If the law of gravitation is strictly the Newtonian law, this leads to an apparent displacement of a star close to the sun's limb amounting to 0″·87 outwards.

(3) The course of a ray of light is in accordance with EINSTEIN's generalised relativity theory. This leads to an apparent displacement of a star at the limb amounting to 1″·75 outwards.

Figure 7.3. A scientific paper: the opening of Eddington's report on the eclipse. (Because the Astronomer Royal Frank Dyson was the nominal head of the project, his name appears first.)

almost exactly what Newtonian physics predicts. Again, no one disputes that the number is calculated fairly and correctly.

There are outbreaks of objectivity throughout the eclipse paper, then. What about subjectivity? Eddington and his team, as you know, chose to put aside the Brazilian astrographic data, which allowed them to draw their dramatic conclusion:

No.	Names.	Photog. Mag.	Co-ordinates. Unit = 50'.		Gravitational displacement.			
					Sobral.		Principe.	
			x.	y.	x.	y.	x.	y.
		m.			"	"	"	"
1	B.D., 21°, 641	7·0	+0·026	−0·200	−1·31	+0·20	−1·04	+0·09
2	Piazzi, IV, 82	5·8	+1·079	−0·328	+0·85	−0·09	+1·02	−0·16
3	κ² Tauri	5·5	+0·348	+0·360	−0·12	+0·87	−0·28	+0·81
4	κ¹ Tauri	4·5	+0·334	+0·472	−0·10	+0·73	−0·21	+0·70
5	Piazzi, IV, 61	6·0	−0·160	−1·107	−0·31	−0·43	−0·31	−0·38
6	υ Tauri	4·5	+0·587	+1·099	+0·04	+0·40	+0·01	+0·41
7	B.D., 20°, 741	7·0	−0·707	−0·864	−0·38	−0·20	−0·35	−0·17
8	B.D., 20°, 740	7·0	−0·727	−1·040	−0·33	−0·22	−0·29	−0·20
9	Piazzi, IV, 53	7·0	−0·483	−1·303	−0·26	−0·30	−0·26	−0·27
10	72 Tauri	5·5	+0·860	+1·321	+0·09	+0·32	+0·07	+0·34
11	66 Tauri	5·3	−1·261	−0·160	−0·32	+0·02	−0·30	+0·01
12	53 Tauri	5·5	−1·311	−0·918	−0·28	−0·10	−0·26	−0·09
13	B.D., 22°, 688	8·0	+0·089	+1·007	−0·17	+0·40	−0·14	+0·39

Figure 7.4. Objectivity epitomized: the names of stars surrounding the eclipsed sun, their positions (under "Co-ordinates"), and—under "Gravitational displacement"—Einstein's predictions for the apparent shift in their positions, when viewed both from the Brazilian location (Sobral) and from the African island of Príncipe.

Thus the results of the expeditions to Sobral [Brazil] and Príncipe can leave little doubt that a deflection of light takes place in the neighborhood of the sun and that it is of the amount demanded by Einstein's generalized theory of relativity.

The subjectivity in the paper must manifest itself, then, in Eddington's argument that the Brazilian astrographic was functioning so badly that its measurements should in effect be ignored.

Comb the paper looking for that argument, however, and you'll come up empty-handed. It is nowhere to be found. Eddington does tell his readers that the astrographic images were "diffused and apparently out of focus." He then speculates on the cause ("the unequal expansion of the [telescope's external] mirror through the sun's heat"), and he is very clear that the images are to be given "much less weight"—indeed, by the end of the paper, apparently no weight whatsoever.

But there is something essential missing from this chain of reasoning. Even if Eddington is correct that the "diffusion" is caused by the distortion of the mirror, it is, as he acknowledges, "difficult to say whether this caused a real change of scale in the resulting photographs or merely blurred the images." A change of scale would result in a systematic error, possibly yielding a misleadingly low value for the bending angle, but mere blurring would not. To justify his disregard for the Brazilian photos, then, Eddington would have to share his reasons for believing that there was in fact a scale change. He does nothing of the sort.

The eclipse paper is, as a consequence, peculiarly deficient as a piece of rhetoric (although Eddington made up for that behind the scenes). At the same time and for the same reason, it maintains not only a semblance of but a genuine claim to objectivity. There is little in it that can be contested: not the measurements, not the calculations, not the observations about the clarity or blurriness of various sets of plates, not the speculations about the cause of the blurriness and its possible consequences (you can hardly argue with a mere speculation), and not the logic that says that data from a broken instrument ought to be given little or no weight. The objectivity is not quite complete—the bare claim that the astrographic telescope yielded no useful information can certainly be disputed—but it is close; certainly, the subjective considerations that we suppose pushed Eddington to draw that conclusion, running from the beauty of relativity to the ugliness of postwar politics, are entirely absent.

It is the iron rule that demanded objectivity of Eddington's paper, as it does of all scientific communications. Everything subjective in a scientific argument, the rule says, must be eliminated. Directly in the firing line are plausibility rankings, scientists' estimates of the likelihoods of various important assumptions. If a plausibility ranking does not reflect a broad scientific consensus or an obvious conclusion from incontrovertible premises, it needs to go. Eddington could not, then, simply say that he was confident that the Brazilian astrographic suffered a change of scale (as we may surmise he believed). He would be permitted to give his reasons for thinking there had been a change of scale, but only if they were based substantially on something other than personal opinion or guesswork. That was not the case here; he had no independent evidence that there was not simply a loss of focus, a "mere blurring." He was therefore forced to leave the argument hanging.

So it is in general: when a scientific paper is written, the grounds of many of the experimenter's crucial assumptions, being partially or wholly subjective, are cut away. What is left are only observation reports, statements of theories and other assumptions, and derivations that connect the two. Consequently, arguments appearing in official scientific venues—such as Eddington's published argument for ignoring the Brazilian astrographic—are characteristically incomplete, perfunctory, or oddly blunted. Strange though it may seem, that is what the iron rule's insistence on objectivity entails.

I call this process in which subjectivity is excised from scientific argument "sterilization" in honor of the great sterilizer Pasteur, who understood that the appearance of objectivity is as important as the real thing. My compact formulations of the iron rule typically do not make the demand for sterilization explicit. But it is there.

THOSE ARCHENEMIES OF the scientific method, the radical subjectivists, do not dispute the existence, indeed the preeminence, of a norm of objectivity in science. But they take it to be pure propaganda. The iron rule may stipulate that scientific argument should be sterilized, cleansed of all subjectivity; the reality of scientific reasoning is, however, the usual human story: bias, contextuality, the whole teeming ecosystem of the human mind with its desires and fears, its antagonisms and allegiances, its need to please and its will to believe.

When the neuroscientists Cajal and Golgi scrutinized the same nerve tissue, each observed just what his own theory of the brain predicted and lambasted the other for gross failures of objectivity. Such pontification seems to amount to little more than "You don't see what I want you to see." The philosopher Karl Popper claimed that evidence could be interpreted using an indisputable rule of logic; in fact, what gets "falsified" turns out to depend on plausibility rankings, subjective estimates of various crucial assumptions' likelihoods.

The iron rule, then, must be a cover-up, a ruse that dresses up as objective something—scientific deliberation—that is anything but. According to the subjectivist philosopher of science Paul Feyerabend:

> There is hardly any difference between the members of a "primitive" tribe who defend their laws because they are the laws of the gods . . . and a rationalist who appeals to "objective" standards, except that the former know what they are doing while the latter does not.

What, according to subjectivists like Feyerabend, is the function of the iron rule? To make scientists feel good. To make the nations that fund science feel good. To make Western civilization feel good at having overcome an atavistic mélange of emotion, partiality, and prejudice to attain an exalted kind of knowledge that is purified of its human origins.

In 1936, Stalin oversaw the introduction of a new constitution in the USSR, guaranteeing freedom of speech, freedom of the press, freedom of assembly, and freedom to demonstrate in the streets. By 1937, his regime was executing perhaps one thousand people every day. Sometimes the finest words are nothing more than words. So it is, according to the more radical of the subjectivists, with the iron rule. It is the Soviet constitution, the Marlboro Man, *Arbeit macht frei*.

The subjectivists see the moral bankruptcy of the iron rule, the gulf between science's ideology and its reality, in two apparently contradictory sentences:

The ideology: Scientific argument deals only with the objective implications of empirical tests.

The reality: Scientific reasoning relies essentially on the subjective interpretation of empirical tests.

But there is no contradiction and little conflict. As far as the iron rule is concerned, *argument* and *reasoning* are two quite separate things.

Reasoning is what scientists do in their heads to get from the outcomes of tests to opinions, convictions, and plans of action. It is how they make up their minds whether some theory is surely false, likely true, or still up in the air. It is how they decide whether some research program is staid, foolish, or risky but bold. It is how they determine for themselves whether a therapy or experimental procedure is reliable, hopeless, or simply unproved. Vital to such thinking are plausibility rankings, which supply the auxiliary assumptions on which all scientific reasoning is based. It is because plausibility rankings are essentially subjective that scientific reasoning is essentially subjective.

Scientific argument, by contrast, in the sense that matters to the iron rule, is what appears in science's official channels for broadcasting research, namely, the scientific journals and conference presentations. It is only in such venues, on the printed page and the projected slide, that objectivity is required.

There are, of course, many other places in which scientists argue: in their lab meetings, in their conversations over a beer after a hard day's research, and in their public but unofficial communications, such as television interviews, popular books, and talks at public libraries and corporate retreats. The iron rule does not require objectivity in any of these endeavors. When I say that the rule is concerned exclusively with scientific argument, then, I use that term in a narrow sense—shorthand, in effect, for "official scientific argument."

Official scientific argument excludes not only scientists' informal public communication but also their private thoughts. That resolves the tension between the essential subjectivity of science and the iron rule. Reasoning and argument coexist peacefully in science: one in the minds of scientists, drawing on their judgments, feelings, and inclinations; the other in science's designated organs of communication, which because of sterilization is quite devoid of such things.

A call for objectivity is not in itself unique to modern science; mathematical inquiry, for example, makes the same demand. What distinguishes science from other like-minded pursuits is, first, that it asks for objectivity only in official publications and not in private reasoning, and second, that in pursuit of the objective ideal, it will dismember an argument and throw away essential parts, just as Eddington omitted from his paper any reason to think that the distortion of the astrographic telescope's mirror led to a change of scale. Not objectivity, then, but the singular kind of objectivity achieved by sterilization, is special to science.

THE SUBJECTIVIST CRITIQUE of scientific objectivity as propaganda supposes that the primary function of a scientific paper is to assemble an argument, to make a case, to provide reasons to accept or to reject a hypothesis. The assumption is natural enough: scientific papers often take on the form of arguments, and indeed I have called this form of communication "scientific argument." But it is not at all correct. Scientific papers have two important functions. Neither requires that they articulate genuine arguments. Both benefit immensely from the process of sterilization.

One function of scientific papers is to serve as moves in the scientific game. Sterilization, by putting restrictions on the form that these moves can take, helps to make the rules of the game as clear and simple as possible, thereby bolstering modern science's procedural consensus. A sterilized paper contains only the outcomes of empirical tests and demonstrations that some theoretical cohort or other either explains or fails to explain those outcomes (or ascribes to them a certain probability). The shallow causal conception of explanation, known to and agreed upon by all scientists, determines what satisfies these criteria. Thus, the rules for writing a legitimate paper are as plain as the rules of the game of chess. Every scientist can trust that even though their colleagues may question the quality of their moves, they will not challenge their fundamental permissibility. Secure in this knowledge, they may throw themselves wholly, utterly into the contest.

The other function of scientific papers is archival: they constitute a permanent record of scientists' empirical testing. Sterilization puts these archives in a codified form that is useful to readers who may share few of the preconceptions of a paper's author. A sterilized paper, containing only the outcomes of empirical tests and the explanatory relations between those outcomes and one or more theoretical cohorts, is like a construction kit that other scientists can use to build their own arguments. To the eclipse paper, for example, add a high plausibility ranking for the assumption that the Brazilian astrographic suffered a change of scale, underreporting the bending angle, and you have a powerful argument for Einstein's theory of relativity. Add, by contrast, a high plausibility ranking for the telescope's suffering only from a lack of focus, and you will conclude that it provided useful if unrefined information suggesting a Newtonian bending angle, thus yielding overall an ambivalent case for Einstein, with different sets of measurements pulling in different directions.

Every reader, then, pours their own plausibility rankings onto the desiccated framework of a scientific article, bringing it to life and drawing conclusions accordingly—conclusions that are saturated with subjectivity and that consequently differ from scientist to scientist. Each epoch, each research group, each individual scientist will interpret the scientific literature, and so the significance of the amassed scientific evidence, in their own way.

Early on there will be doubt and disagreement, but as the evidence accumulates, it will emerge that whatever plausibility rankings you bring to the findings, one theory is far better than the rest at explaining all that has been observed. The meticulous preservation of data in its sterilized form smooths and speeds this process of Baconian convergence.

The archival function of sterilized papers is, in short, to capture evidence in a kind of presupposition-free suspended animation ready for the use of future generations. You'd rather read a novel than the labels on carefully organized freezer bags. That is why scientific writing—in journals and conference proceedings—is as dull and respectable as writing about the inner workings of science is delectable.

THE SCIENTIFIC JOURNAL in which Eddington's eclipse report was published—the *Philosophical Transactions of the Royal Society*—was established in 1665 and is generally considered the oldest publication of its kind. From the very beginning, the *Philosophical Transactions* took on the character of an archive of objective fact, a "vast pile of experiments," in the words of Thomas Sprat, one of the first members of the Royal Society, who in 1667 characterized its purpose thus:

> The Society has reduced its principal observations, into one common stock, and laid them up in public registers, to be nakedly transmitted to the next generation of men.

That official, public scientific writing should be concerned exclusively with naked fact, stripped of the opinions and untainted by the aims of the author—that is the ideal of sterilization, the "drive to objectivity" of this chapter's title, present as you can now see from science's earliest days.

The rhetoric of objectivity has evolved. During Sprat's time—the first decades of the Royal Society's existence—authors emphasized their presence in the laboratory, using an active, first-person narrative to communicate what was seen to occur. They painted the scene with numerous details of only passing relevance, striving to give their readers a sense that they themselves were there alongside the experimenter, seeing the phenomena with their own eyes. In a characteristic report, the physicist Robert Boyle wrote:

> We took a slender and very curiously blown cylinder of glass, of nearly three foot in length, and whose bore had in diameter a quarter of an inch, wanting a hair's breadth; this pipe, being hermetically sealed at one end was, at the other, filled with quicksilver . . .

I spare you the rest. As the historian Peter Dear, who quotes this passage, aptly observes, such writing conveys with great force "the actuality of a discrete event."

A little of this persists in Eddington's eclipse report from one hundred years ago. It relates a number of inessential narrative details, such as the names of the ships on which Eddington and his collaborators sailed and the island of Príncipe's dominant vegetation and prevailing climate ("very moist, but not unhealthy"). Contemporary scientific writing, by contrast, proclaims its objectivity through its ruthless exclusion of any such elaborations. Its color is gray, its nakedness clinical.

More significant than changes in style over the centuries are changes in the standards for content. The methods of statistical analysis in Eddington's eclipse report, although sound, are somewhat informal and are not consistently applied. More important still, the paper asserts forthrightly that the Brazilian astrographic measurements "are of much less weight" than the others, a remark that amounts to the oblique declaration of a plausibility ranking. Practices vary among disciplines and journals, but you would be far less likely to find these things in a scientific article today. The long-term trend is toward ever more thorough sterilization, toward methods of presentation in which the outcomes of empirical tests are exposed directly to the reader, apparently free of authorial intercession.

New ideas and new technologies have ever held out hope of fully realizing the objective ideal. With the advent of photography in the nineteenth century came the prospect that the subjectivity inherent in using words or drawings to capture what is observed might be overcome. Figure out how to photograph clearly the fine structure of the brain, for example, and Cajal's and Golgi's feuding becomes beside the point; everyone will see for themselves how neurons connect. Surely a camera shows the world as it really is, independently of the scientist's mind or eye?

At the same time, the development of formal statistical methods promised unblemished objectivity in argument. A simple mathematical formula would take the raw data as input and supply as output, without any human interference or evaluation, a judgment as to whether or not a scientific claim was credible enough to be published—whether it would be acceptable to assert in scientific print that smoking causes cancer, that the Higgs boson exists, that human activity is causing global warming.

Neither of these promises has been entirely fulfilled. Even the most objective statistical techniques leave some choices up to the scientist, and these choices, it has become increasingly clear, can be gamed to illuminate the data from the most favorable (or publishable) angle. As for photographs, the obstacles to achieving the objective ideal are crisply illustrated by the scientific study of snowflakes.

To discern the geometry of a snowflake, you need a microscope and good cold temperatures—the canonical six-sided flake forms only below –15 degrees Celsius, or 5 degrees Fahrenheit. Work on examining and depicting snowflakes began seriously in the seventeenth century in Robert Hooke's famous compendium of microscopic observations and continued through the next two hundred years as better microscopes

Figure 7.5. Snowflake forms drawn by Robert Hooke (1665), left; by William Scoresby (1820), right.

resulted in ever more intricate illustrations of their six-sided symmetry (Figure 7.5).

Whereas Hooke's flakes are a little uncouth, the resplendent forms drawn by the Arctic explorer William Scoresby in 1820 exhibit exemplary symmetry. Scoresby saw God imparting his own perfection to his creation:

> The particular and endless modifications of similar classes of crystals, can only be referred to the will and pleasure of the Great First Cause, whose works, even the most minute and evanescent . . . are altogether admirable.

Are snowflakes really quite so immaculate? Scoresby drew them that way, but other snowflake scientists' reports might prompt second thoughts.

During one particularly brutal winter freeze in 1855, the British meteorologist James Glaisher sketched snowflakes and handed them on to an illustrator to provide the finishing touches. The drawings were lovely, each flake flawless in its own six-sided way. As Glaisher remarked, however, the ideal was not perceived but rather inferred: the sketches had none of it. The illustrator had assumed the symmetry of the flakes in order to round out the finished pictures. That a drawing is made from life, then, does not preclude its being permeated by presupposition; in this case, the regular proportions represented on the page reflected the illustrator's convictions better than they reflected the world.

Photography promised to eliminate these subjectivities. There would be no interpretation, no opportunity for artistic idealization, only light traveling directly from flake to photographic plate to paper to the reader's eye.

The Vermont farmer for the job was Wilson Bentley. Extreme cold, a bellows camera, his mother's microscope, and a finely etched aesthetic sense—these were all he needed to begin to capture and photograph snowflakes in 1885 at the age of 19. By the time of his death from pneumonia in 1931, he had framed thousands of images; of these, over 2,000 were to be published in his book *Snow Crystals*. Figure 7.6 (left) shows a sample of his work; as you can see, the flakes are as perfectly proportioned, as symmetrical, as in Scoresby's drawings.

Such perfection made the German meteorologist Gustav Hellmann angry. Hellmann had spent years reconstructing the structure of snowflakes by hand from brief glimpses under a microscope, like Glaisher's assistant using the principle of symmetry to fill out missing parts. Eventually, he turned to the new technology of photography: in 1892, he and his collaborator Richard Neuhauss began to make images of flakes using an advanced camera. What they saw was quite different from the forms captured in Bentley's photos. Hellmann and Neuhauss's flakes had their own rough beauty, but perfect symmetry was vanishingly rare (Figure 7.6, right). Bentley, they concluded, had

Figure 7.6. Snowflakes photographed by Wilson Bentley (1901), on the left; on the right, snowflakes photographed by Doug and Mike Starn, showing the pervasive irregularities observed by Hellmann and Neuhauss.

touched up his images to "correct" their imperfections. Worse, Neuhauss later wrote:

> In many images Bentley did not limit himself to "improving" the outlines; he let his knife play deep inside the heart of the crystals, so that arbitrary figures emerged.

Bentley's crystals were, in other words, ice sculptures of his own design.

The feud continued for decades. We don't know whether Neuhauss's worst suspicions were correct, but it seems clear that, at the very least, Bentley was not averse to a little post-processing, "the old-school version of Photoshop," as contemporary snowflake photographer Ken Libbrecht, a professor of physics at Caltech, has put it. Certainly, Bentley's depiction of the natural world is rather misleading; Libbrecht has found that only one in a thousand snowflakes has perfect six-sided symmetry. Even photography, it seems, allows space for a scientist's editorial inclinations; even the camera has a point of view.

The story of snowflakes shows as well as any that the iron rule's ideal of objectivity for arguments appearing in official scientific communications is only ever partially achieved. But it is not merely a pretense, not simply propaganda. The presentation of scientific evidence and argument may fall short of perfect objectivity while still largely performing its underlying functions: to archive for future generations the outcomes of tests and their explanatory relations to theory and to channel scientists' energy and attention away from opinion, persuasion, and invective—directing it instead toward the production of exquisitely detailed empirical fact.

CHAPTER 8

The Supremacy of Observation

℘

How the iron rule ejects everything from scientific debate but a
theory's ability to explain observable phenomena—and how
that came to be (the iron rule's fourth innovation)

*O*NLY EMPIRICAL TESTING COUNTS. Across the world of science, this thought is expressed over and over in some form or other. "Experiment is the sole judge of scientific truth" (physicist Richard Feynman). "Take nobody's word for it" (motto of the Royal Society). "To experience we refer, as the only ground of all physical enquiry" (nineteenth-century man of science John Herschel). "Observation is the generative act in scientific discovery" (biologist Peter Medawar). "All I'm concerned with is that the theory should predict the results of measurements" (physicist Stephen Hawking). Such words are an invocation and a celebration of modern science's fourth great innovation, its fourth great departure from the empirical inquiry of old: a prohibition on any form of persuasion, however well founded, however objective, that is not based on empirical testing.

To explain how the prohibition works and to get a sense of its novelty and peculiarity, we will visit Britain in the 1830s, where it is already at full operational capacity, shutting down any attempt to import philosophical or religious thinking into scientific argument. From there we will travel more than a century further back in time, to catch the iron rule's censorious impulse at the moment that it takes control—thereby witnessing a critical moment in the creation of modern science and perhaps the most consequential invention in the history of thought.

WILLIAM WHEWELL DIED after falling from his horse in 1866. He was 71 years old and left behind him a glittering trail of achievements. Born the son of a carpenter, he attended Trinity College at the University of Cambridge as a "subsizar," paying reduced fees in return for waiting on the tables at which the other students and the Fellows of the college took their meals. He won prizes for poetry and mathematics, became a Fellow at Trinity himself, was appointed a professor of mineralogy and then moral philosophy, and in 1841—29 years after he first entered Cambridge as a table-waiting scholarship student—he was elected Master of Trinity College.

Whewell's long-distance leap up the social ladder had its costs. Surrounded by his social "betters," he was sensitive and proud, quick to take offense. Moving from the local school for working-class children to a more elite grammar school, he soon found himself using his considerable physical size to fend off his bullying classmates; later on his friend John Herschel remarked that Whewell's "temper will never be good!"

Though master of a great college, he was socially awkward. According to the biographer Leslie Stephen:

> In early days he had little chance of acquiring social refinement; and, though he was anxious to be hospitable, his sense of the dignity of his position led to a formality which made the drawing-room of the lodge anything but a place of easy sociability.

What he lacked in charm, however, he made up for in industry.

At Trinity, he instituted a program of educational reform while writing about almost everything: the tides, astronomy, Gothic architecture, theology, mechanics, ethics, and the probability of life on other planets. He translated Goethe and Plato. He invented the self-registering anemometer, a device to determine wind speed. He descended a mine in Cornwall to measure the earth's gravitational field; he climbed mountains in Switzerland to contemplate the glory of God in the empyrean heights. And he coined a new word for the thinkers who had until his time been known as "men of science" or "natural philosophers": he proposed that they be called *scientists*.

Of all this prolific output, Whewell's greatest achievement was a three-volume history of the sciences, published in 1837, followed in 1840 by a two-volume philosophical study of their method, which as the title proclaimed was "founded upon their history"—a monumental pentalogy that makes Whewell one of the earliest and also one of the most distinguished historians and philosophers of modern science.

The last of the sciences to be considered in Whewell's history was the young discipline of geology. Whewell rather unconventionally included under the heading of geology the history of life, a story that had begun to emerge only in the previous 50 years in the layers of fossilized shells and bones uncovered by the Industrial Revolution's canal diggers and railway builders. The most startling chapter in this piecemeal chronicle revealed that long before the appearance of large mammals, the earth was dominated by dinosaurs and their kin. As the fossil-hunting doctor Gideon Mantell colorfully put it in "The Geological Age of Reptiles" (1831):

There was a period when the earth was peopled by [egg-laying] quadrupeds of a most appalling magnitude, and . . . reptiles were the *Lords of the Creation*, before the existence of the human race!

Figure 8.1. England in the Age of Reptiles: George Nibbs's imaginative visualization of *The Ancient Weald of Sussex* (1838). At the center is an early and highly inaccurate reconstruction of the iguanodon, a dinosaur that Gideon Mantell was the first to describe.

That was only one of the epochs during which the planet had evidently been populated by wholly different life-forms; before the dinosaurs there were seas roiling with strange marine creatures—trilobites, ammonites, and primitive jawless fishes—and after the reptiles came pygmy horses and giant sloths.

To Whewell, the lesson taught by the rocks was plain. "The species of plants and animals which are found imbedded in the strata of the earth," he wrote, "are . . . different from any now existing on the face of the earth. . . . They imply . . . that the whole organic creation has been renewed, and that this renewal has taken place several times." How, Whewell asked himself, were these episodes of renewal—such as the replacement of the dinosaurs by mammals—brought about? New forms of life, he answered, are created by "other powers than those to which we refer natural events," or in other words, by the Christian God.

When Whewell's history was published in 1837, such an explanation was both reasonable and conventional. Charles Darwin was only 28 years old, freshly returned from his voyage around the world on the HMS *Beagle*. His masterwork, *On the Origin of Species*, would not emerge for another 20 years. There was little reason, then, for Whewell and his contemporaries to doubt that, just as the Bible said, "out of the ground the Lord God formed every beast of the field and every fowl of the air" (Genesis 2:19).

What was new was incontrovertible evidence that God had had more than one go at it. The creation of life had happened not only in some unknowable moment at the very beginning of time, but again and again, with the contours of each iteration impressed unmistakably in the fossil record.

That opened up an extraordinary opportunity to pursue two grand projects. First, knowledge of patterns in the rocks corresponding to episodes of renewed creation, available to anyone with a trained eye and a geologist's hammer, could cast light on God's mind, on his intentions and plans in fashioning this planetary receptacle for his supreme invention, the human race. And second, knowledge of God's intentions and plans, available to anyone with access to the holy scripture and a theological turn of mind, could cast light on the natural history of the world, on the marching order of life's grand parade.

As to the first project, some thinkers might have recoiled at the impiety, or at any rate the sheer temerity, of using scientific knowledge to plumb the divine intellect. Not Whewell. When the eighth Earl of Bridgewater, a clergyman, naturalist, and antiquarian, set aside a bequest to sponsor the publication of works devoted to finding signs of God's plan in nature, Whewell eagerly signed on. In 1833, he contributed a volume to what became known as the Bridgewater Treatises, in which he contended that "every advance in our knowledge of the universe harmonizes with the belief of a most wise and good God." From the configuration of the planets, the arrangement of the earth's surface, and the sophisticated design of living things, we can see—according to Whewell—that "the wise and benevolent Creator of the physical world" is also the "just and holy" "Governor of the moral world" and thus the "Judge of men."

Writing his history and philosophy of the sciences in the late 1830s, then, Whewell would seem to be the ideal man also to lead the second project, which at that moment in time exhibited such promise: the integration of geological and theological knowledge to give a complete account of the history of life on earth. He had the empirical expertise, the religious motivation, and a preeminent position in the world of science.

But he did not do it. Indeed, in his *History of the Inductive Sciences*, he is adamant that no one should do it and more generally that science should take no account whatsoever of theology, which "must never be allowed to influence our physics or our geology." Biologists and geologists should attend only to "the ordinary evidence of science"—that is, to the outcomes of empirical observations—and should go only as far as that evidence can carry them. Theology, meanwhile, can use its proprietary resources, such as philosophical reasoning about the nature of God and scriptural inter-pretation, to make its own inferences about life's creation. When both endeavors conclude their investigations, their findings can be merged to provide a complete picture of the history of life—but not before.

Whewell's embargo on the second grand project, the importation of theology into geology, was not prompted by concerns about theology's irrelevance or impracticality. As best he could determine, the project was feasible and likely fruitful. The geologist's goal was to set out the course and the causes of the unfolding history of life, and in Whewell's view, the most important turning points in that history were effected by the hand of God. Without the assistance of theology there would be, at each of these pivotal moments, an empty space in the scientific annals of the earth's past. Further, Whewell believed that theology was able to fill these blanks: in the course of his Bridgewater Treatise he argued that God's nature and plans manifested themselves in every aspect of the natural world. He repudiated the project all the same.

It is a conundrum. There is a need, the means to satisfy that need—yet in Whewell's prescription of methodological apartheid, a stern injunction against doing so.

The injunction makes no cultural sense: in the 1830s and 1840s, virtually every scientist was a believer, attributing the existence of the universe and all its marvelous peculiarities to the machinations of the Creator. As the Bridgewater enterprise shows, the zeitgeist made ample room for a merging of natural history and religion.

Nor does it make logical sense. Two projects working toward the same end should always be open to collaboration. Imagine two children searching for a lost dog. One goes from house to house, knocking on doors and asking if there have been any sightings. The other roams the streets, parks, and junkyards calling the dog's name. They must separate in order to pursue their distinct strategies. But it is clearly in their interests to check in with one another from time to time: if the dog has been seen on the east side of town, better to explore the parks on the east side than the west side. What Whewell's separation of theology and geology required, however, was two completely disconnected searches.

Finally, intellectual partition makes, for Whewell, no emotional or psychological sense. It is contrary to his synthesizing spirit, captured delightfully in his words to a friend about a forthcoming visit to the English Lake District:

> You have no idea of the variety of different uses to which I shall turn a mountain. After perhaps sketching it from the bottom I shall climb to the top and measure its height by the barometer, knock off a piece of rock with a geological hammer to see what it is made of, and then evolve some quotation from Wordsworth into the still air above it.

Whewell believed that thought and action were to be woven through the world from multifarious directions but with a single thread. The exclusion of God from geology was, however, a proposal to cut the thread, indeed, to tear up the tapestry into motley, ragged patches of knowledge.

No emotional sense, no cultural sense, no logical sense. Yet what Whewell prescribed is a straightforward application of the iron rule—more exactly, of that aspect of the rule that says *only empirical testing counts*. When Whewell insisted on admitting to geology only empirical considerations, only rocks, fossils, clues chiseled out of the earth, he was doing what the iron rule, already firmly entrenched in scientific practice, told him to do. When his fellow scientists accepted his proposal without outrage, concern, or even vigorous comment, considering the exclusion of religious thinking from geology to be as congenial and conventional as would any twenty-first-century scientist, they, too, were guided by the rule.

Many contemporary scientists believe that religion has little or nothing that is useful to say about the history of life on this planet. Whewell and his peers, by contrast, took their religious knowledge to be both trustworthy and biologically revealing. That is why it is so illuminating to see the iron rule at work in 1837, why I have made the effort to take you back to Whewell's time rather than merely examining the practice of science in our own day. Whewell did not invent the iron rule—he was born far too late for that—but in his efforts to reconcile his scientific research with his religious belief, he casts its strictures in stark relief.

Under this revealing light, it is apparent that the iron rule excludes from scientific argument more than just "subjective" reasons. It excludes every nonempirical consideration, no matter how persuasive or well founded. To Whewell, the existence of the Christian God was as clear and certain as any observable fact. Yet he accepts that theological considerations ought nevertheless to be disbarred from official scientific dispute in accordance with the rule—not because they are bad reasons, not because they are purely personal reasons, not because of any perceived logical defect, but simply because they are the wrong kinds of reasons for doing science.

The iron rule, then, legislates a distinction between scientific and unscientific reasons that is not at all the same as the distinction between objective and subjective reasons, or between strong and weak reasons, or between good and bad reasons. Scientific reasons to endorse a theory are supposed to be objective, strong, and good, but that is not enough: even the most powerful argument is excluded from science unless it is empirical, that is, grounded in a theory's ability to explain observed fact.

Whewell's case illustrates something else about "only empirical testing counts" that is central to the iron rule's operation: it applies to public scientific argument but not to private scientific reasoning, and even then only to argument in science's institutionally sanctioned venues for debate. The rule in no way prevented Whewell from musing to himself about the biological significance of his religion. It merely prevented him from exploring it in scholarly publications whose primary purpose was scientific communication. In his Bridgewater Treatise, then, a piece of popular theological writing, Whewell was quite free to speculate about God's part in the history of life.

Informal outlets for scientifically inspired religious expression have not vanished. Kelvin kept God out of his official publications about the age of the earth, but in an 1889 talk to the Christian Evidence Society, he felt free to follow Whewell in discerning, in the scientific world picture, clear evidence for the existence of a creator. And Francis Collins, the leader of the Human Genome Project, maintained the tradition in his 2006 book *The Language of God: A Scientist Presents Evidence for Belief.*

The iron rule ignores all such productions—it ignores anything outside science's anointed journals and conference proceedings—because they do not constitute moves in the game of modern science. To regulate that game is the rule's sole concern. Its first three great innovations— shallow explanation, the demand for objectivity, and the distinction between reasoning and official argument—work together to ensure that all scientists share an understanding of what makes for an empirical move in the game. The fourth and final innovation, the "only" in "only empirical testing counts," insists that every move be empirical and in so doing transforms the game's players into observational and experimental prodigies, into extractors of evidence par excellence.

Whewell believed that theology was an essential source of knowledge about the natural world. Nevertheless, he wanted to play the game. Theology, therefore, and all the insight that from his point of view stood to come with it, was abandoned on the sidelines.

The pain caused by that desertion shattered Whewell's customary clarity of thought. Arguing against any effort to unite theological and geological reasoning, he wrote that to attempt such a synthesis would be to assume

> that reason, whether finite or infinite, must be consistent with itself; and that, therefore, the finite must be able to comprehend the infinite, to travel from any one province of the moral and material universe to any other, to trace their bearing, and to connect their boundaries.

Perhaps this struck some of his readers as profound. To my philosophical ear, it is pure gobbledygook. Unable to reconcile the iron rule's directives with his intellectual conscience, Whewell was floundering; he was back in the schoolyard, trying to brazen his way out of trouble by sheer bluster.

Whewell's suffering is a clue to a crucial fact: for an intellect open to every respectable reason, to every relevant consideration, to every good argument, the now familiar doctrine that "only empirical testing counts" is, in its illiberality, quite alien. That unsettling doctrine was already firmly in place when Whewell embarked on his scientific career. How did it come to govern all deliberation about the structure of the natural world? What was the origin of this most striking and important aspect of the iron rule? To find the answer, we can stay right where we are, in Trinity College, Cambridge, but we must turn back the clock another 150 years.

ARRIVING IN THE LATE SPRING of 1681, you might, with some agility and a certain disregard for the regulations, find your way into a private, walled garden to the right of the Great Gate at Trinity College. Suppose you do it.

In the garden you come upon a wooden structure built against a wall, and in that structure a well-provisioned laboratory. A notebook sits on the lab bench. Opening it at random, you read the following curious formulation:

> Neptune with his trident leads the Philosopher into the sophic garden.

There is no trident to be seen, and certainly no sea god—just furnaces, crucibles, alembics, and a welter of nameless substances. About the identity of the Philosopher, however, there can be no doubt, for you are standing in the laboratory of Isaac Newton, inventor of the calculus, discoverer of the laws of motion and gravity, analyzer of light—and one of the greatest alchemists of his age.

Deciphering page after page of Newton's densely written text in the waning light of the afternoon, you find a tantalizing description of the celebrated "green lion":

> Concerning Magnesia or the Green Lion. It is called Prometheus
> & the Chameleon. Also Androgyne, and virgin verdant earth in

Figure 8.2. Newton's laboratory at Trinity College, Cambridge, was located somewhere within his private garden (bottom right), reached by a covered staircase from his chamber on the middle level just to the viewer's right of Trinity's Great Gate. Newton lived in this chamber from 1673 until he left Trinity in 1696. From a 1690 engraving by David Loggan.

which the Sun has never cast its rays although he is its father and the moon its mother.

You turn to the end of the notebook. There lies recorded an astonishing discovery:

May 10, 1681. I understood that the morning star is Venus and that she is daughter of Saturn and one of the doves . . . May 14. I understood the trident. May 18. I perfected the ideal solution . . . the eagle carries Jupiter up.

What had Newton found out?

Perhaps the answer lies in an undated note describing the extraction of something extraordinary from the veins of the green lion:

> Dissolve volatile green lion in the central salt of Venus and distill. This spirit is . . . the blood of the green lion Venus, the Babylonian Dragon that kills everything with its poison.

Other notes identify the blood of the green lion with "vivified mercury," a substance capable of destroying even gold—"the green lion devouring the sun."

As to the nature of vivified mercury, we are in the dark. (It is not to be confused with the ordinary element.) Likewise, although "the trident" appears to be a kind of chemical process, we have little idea how it might work. Newton's fantastic language is a kind of alchemical code inspired by the language of other alchemists, in which "green lion" refers to stibnite, or antimony ore, "Venus" refers to copper, and "doves of Diana" means silver. His writings cannot, however, be fully understood in terms of these conventional meanings. Clearly, the Babylonian dragon is a more arcane quarry than antimony ore or ordinary mercury—but because Newton did not explain the purpose of his experiments or supply a key to his system, we have no idea what, precisely, he supposed he had discovered in the middle of May 1681.

Whatever were the goals of his alchemical research, Newton pursued them relentlessly: he alchemized for decades after being appointed the Lucasian Professor of Mathematics at Cambridge, writing "well over a million words" on the topic. He kept it all to himself, publishing nothing.

Why was Newton so interested in alchemy? The answer is simple, though it sits uneasily with the public image of Newton as a purely scientific intellect, a harbinger of the Enlightenment. Newton, like Whewell, was interested in *everything*—everything, that is, that might possibly reveal the secret powers underlying the workings of the universe. Thus, he studied the motions of the planets, which disclose the principles of universal gravitation. He conducted alchemical experiments "to liberate the spirit or active virtue of bodies from its encumbering feces." And he followed paths of inquiry that involved no experimentation or observation at all, but simply reading and thinking.

Among these nonempirical pursuits was the philosophical investigation of the nature of space, matter, and motion. To this end, Newton studied the works of the preeminent English metaphysicians of his day—the Cambridge Platonists—and penned a treatise (like the alchemy, never published) to refute Descartes's contention that there can be no empty space. In later writings he denounced another Cartesian idea, that all changes in motion are caused by collision. Bouncing molecules, thought Newton, cannot account for substances such as oil and water failing to mix; their tendency to segregate must rather be explained by a "secret principle of unsociableness." In the same vein, he attempted an account of the vital spirit, the spark of life, apparently the cause of all motion:

> The vital agent diffused through everything in the earth is one and the same. And it is a mercurial spirit, extremely subtle and supremely volatile.

Other unpublished writings from the same period—roughly, the 1670s, after Newton had made many of his discoveries in mathematics and physics—display an intense interest in the proper interpretation of scripture and in biblical prophecy. Newton worked hard to find historical events that corresponded to events foretold in the Bible, he attempted to reconstruct the exact plan of the great Jewish temple in Jerusalem, and he used his biblical chronology along with the words of the prophets to predict a date for the Second Coming, which was apparently to occur some time in the late nineteenth century. As he labored, he found himself drawn ineluctably to the heretical Arian doctrine that Jesus Christ was created by, rather than identical to, the Christian God. Though it endangered his position at Cambridge, he never renounced this belief.

Some historians have, in the light of these endeavors, attributed to Newton a belief in the *prisca sapientia*, or ancient wisdom—a profound knowledge of the nature of things supposedly grasped by sages in ancient times and passed on allusively and allegorically by figures such as Pythagoras and Plato. To decode the wisdom takes the sensibility not only of a physicist but also of a philosopher and a poet. The ancient story in which the god Vulcan surprises his wife Venus in the embrace of Mars and captures the couple in a golden net is, according to Newton and other alchemists, to be understood as a recipe for a fecund substance called "the net," a "hermaphrodite" that combines "the male seed of Mars with the female principle of Venus." Only an investigator equally dextrous with analogy and alembics, a rigorous experimenter and a metaphorical scrutineer, could fully exploit such clues. Newton's writings on physics, philosophy, alchemy, and scriptural interpretation indeed seem like fingers reaching for the *prisca sapientia* in any and every possible way.

Francis Bacon, decades earlier, warned about the corruption of scientific inquiry by "superstition and a dash of theology," in which the understanding is seduced by "fantastic, high-blown, semi-poetical philosophy." "Lofty, high-minded characters," he thought, are especially susceptible to these "idols of the theater," which stage entrancing theoretical, even mystical, tableaux that distract the mind from the plain speech of the observable facts. Breaking into Newton's laboratory, we have caught him in flagrante, engaged in the most un-Baconian behavior imaginable.

The rationale for our unauthorized incursion was to seek out the source of the iron rule's decree that only empirical testing counts and thus the origin of the most important tenet of modern science. We seem to have discovered, however, that Newton was in this respect not one of the first scientists but—as the great economist John Maynard Keynes wrote—"the last of the magicians." Have we traveled too far back in time or to the wrong place? Not at all. Take a closer look at Newton's researches and you will see that this particular magician enacted his enchantments in a rather unusual way.

On the one hand, Newton's interests were as broad as those of any Renaissance wizard. But on the other, the methods he used to pursue those interests were kept quite separate. Undertaking the investigation of the physics of motion, light, and gravity, the sole basis of his reasoning was mathematics, astronomical observation, and experiment. In his studies of "chymistry," or alchemy, experiment was again paramount, but it was now blended with allegorical thinking and an entirely new suite of hypotheses about vital agents, mercurial spirits, and the like. In his metaphysical disputes with Descartes about the nature of matter and space, his method was pure thought, that is, philosophical argumentation. And in his theological investigations of the nature of God and God's plan for the human race, accounting for aspects of history that gravity and vital sparks alone could not explain, Newton's methods were scriptural interpretation—for which he laid down 15 rules—and textual criticism. Throughout, Newton made little or no attempt to bring his conclusions in one area to bear on any other. His intellect operated, in short, as if in accordance with some "secret principle of unsociableness" that prevented his different investigations, in spite of their overlapping subject matter, from coming together to share their secrets.

Thanks to this sequestration of methods, as congenial to Newton as it was repellent to Whewell, the work that made Newton famous— his physics of gravity and light—was pursued entirely in accordance with the iron rule's edict that "only empirical testing counts." Alchemy, theology, and the *prisca sapientia* played no role in these investigations. Newton judged physical theories solely with respect to the observable phenomena that they were able to explain.

Yet he was not following the iron rule. He was not following any methodological doctrine at all. What drove him was pure instinct, a quirk of his psychology that made him, quite unlike his seventeenth-century peers, a natural intellectual compartmentalizer.

When he entered the alchemy lab, he not only put on the alchemist's robe; he also assumed the alchemist's persona, taking on their allusive language, their allegorical style of thought, and their conception of matter and the principles of chemical interactions. In his beliefs, his behavior, and his words, he became the alchemist.

Figure 8.3. The alchemist, as imagined by Joseph Wright of Derby.

Figure 8.4. The mathematical physicist, as envisioned by William Blake in the person of Newton himself.

When he left the lab at daybreak and went back to his investigations of gravity, the robe was left hanging among the phials, funnels, and retorts. He was now wholly the physicist, intent on using the geometrical method to explain patterns of motion and concerned exclusively with the question of which trajectories can be derived from what mathematical laws.

Like a great actor, Newton gave each of his characters full rein, inhabiting rather than overpowering them, pushing their distinctive concerns, obsessions, assumptions, and modes of reasoning to the limit, playing them for all the knowledge they were worth. And so method for Newton was always multifarious: a cabinet of masks, a repertoire of dramatic roles. Each line of investigation took place in a different setting following a different script: alchemy in the laboratory; gravity in the observatory; theology in the hermit's cave; philosophy—of course—in the sophic garden.

In keeping religion and philosophy out of empirical inquiry, Whewell was following the iron rule, by his time a social code recognized and respected by all serious scientists. Newton, by contrast, did not need a rule to tell him that in physics, only empirical testing counts. That kind of narrowness was built directly into his channels of thought.

Or at least, such was the case for the Newton who lived and worked at Trinity College, Cambridge, from 1667 to 1696. Years later, he evidently came to see how much his success had depended on his compartmentalizing ways, and at the age of 70, having served for almost two decades as Warden and then Master of the Royal Mint in London, he formulated his instinctive mental habit as a methodological principle, broadcasting it to the world in the postscript to the second edition of the *Principia* (1713):

> Whatever is not deduced from the phenomena must be called a hypothesis; and hypotheses . . . have no place in experimental philosophy.

With these words, he wrought a great half-truth. As a summary of the methods of Newton the man, what he said was pure myth. In his alchemy he sought to read mystical symbols, and in his inquiry into space and time he philosophized vigorously with Descartes's ghost. But as a summary of the method of Newton the mathematical physicist—a dramatic persona rather than a living person, but nonetheless the true author of the *Principia*—what he said was quite accurate. The insights of that extraordinary work were achieved through a process that we would now recognize as an embodiment of the iron rule and in particular of its injunction to attend to no other virtues of a theory than its ability to explain the phenomena. So in the *Principia*'s postscript, Newton gave his successors, and the world, the prohibition on religious, philosophical, and other nonempirical argument in science that constitutes the iron rule's fourth great innovation.

That prohibition, like the iron rule's more positive aspect—the definition of empirical testing in terms of shallow causal explanation—shimmered and sparkled with the Newtonian aura. It was seen by the thinkers who came after Newton as the vital spirit of empirical discovery. Adherence to both aspects of the rule swept across Europe, precipitating a revolution—the Scientific Revolution—that turned the creaking, antique apparatus of natural philosophy into the sleek knowledge-making machinery that is modern science.

NEWTON DID NOT FIGHT the Scientific Revolution single-handed: a number of other leading figures of seventeenth-century natural philosophy helped to prepare the way for the adoption of the iron rule. Although it was Newton's influence that was decisive, these lieutenants deserve a place in any account of the revolution's success. I will pause to read a few names from the roll of honor.

Something approximating the iron rule's prohibition on subjectivity and nonempirical argument can be found, as intimated above, in Bacon's prospectus for science, with its rejection of the "idols" of human nature, language, and culture in favor of the probative power of simple observation. Bacon also anticipated the shallow notion of explanation, criticizing earlier natural philosophers in characteristically unsparing terms:

> It is no less of a problem that in their philosophies and observations
> they waste their efforts on investigating and treating the principles
> of things and the ultimate causes of nature, since all utility and
> opportunity for application lies in the intermediate causes.

Bacon's rationale for shallow explanation is, unlike Newton's, facile in a moral as well as a metaphysical sense: once you have a good set of causal principles, he seems to be saying, your learning the underlying mechanisms by which they operate won't make you any richer. Nevertheless, he must receive a share of the credit for articulating both sides of the iron rule.

The extraordinarily precise naked-eye astronomical measurements of the Danish astronomer Tycho Brahe (1546–1601) were crucial in helping his assistant Johannes Kepler (1571–1630) to formulate his laws of planetary motion, physical principles that were in turn explained by Newton's gravitational theory. Kepler's use of Tycho's observations demonstrated early on the scientific value of minute details of no intrinsic philosophical interest—the immense importance of the digits far to the right of the decimal point celebrated by the Tychonic principle for which I have appropriated Tycho's name.

Robert Boyle (1627–1691), who investigated the properties of gases and numerous other phenomena, proclaimed the virtues of making observations independently of any theoretical conjecture:

> To keep my judgement as unprepossessed as might be with any of the modern theories of philosophy, till I were provided of experiments to help me to judge of them, I had purposely refrained from acquainting myself thoroughly with the entire system of either the Atomical, or the Cartesian, or any other whether new or revived philosophy.

Boyle claims, then, to have proceeded straight to the lab without reading any of the great natural philosophers of the age; when he came to advocate the atomic hypothesis, as he does in the essay from which the preceding passage is taken, it is supposedly on the basis of raw data alone. This theoretical agnosticism is too extreme to serve as a general recipe for science, and in any case modern historians doubt that Boyle's account of his method is accurate. Still, like Bacon in the *New Organon*, he articulated an ideal in roughly the spirit of the iron rule.

Among Galileo Galilei's (1564–1642) many contributions to the modern world picture were his efforts to write down mathematical formulas describing the principles of physics and to derive from them the motions of particular kinds of objects. In *Two New Sciences* (1638), for example, he uses his physical principles to calculate the trajectories of cannonballs and the like, showing that such projectiles will follow a classical mathematical curve—the parabola. This mathematical systematization was central to demonstrating the power of shallow explanation in physics (although, I should add, shallow explanation need not be mathematical). The Dutch thinker Christiaan Huygens (1629–1695) continued the tradition, formulating a physics of collisions and deriving the mathematics describing the motion of a pendulum.

The last scientific revolutionary to be commended here is not a person but an organization: England's Royal Society (founded in 1660 and still thriving), which devised the idea of an objective register of empirical observation, the scientific journal in its embryonic form.

Bacon's prescriptions; the exacting measurements of Tycho Brahe; Robert Boyle's celebration of experiment over theory; Kepler's mathematical astronomy and Galileo's and Huygens's mathematical mechanics; the publications of the Royal Society: all of these contributed to the conception of purely observation-based inquiry codified by the iron rule. The script for playing the empirical scientist, and in particular the mathematical physicist, had by Newton's time been drafted already by Bacon and the rest. None of these thinkers, however, inhabited the role like Newton. He needed no direction. Simply by following his compartmentalizing heart, he acted out the iron rule more stringently, more perfectly, than anyone who came before.

I HAVE DISMANTLED the iron rule, over the last few chapters, to analyze each of its four innovations. Now I need to reassemble the parts, showing how the innovations work together to power and steer the knowledge machine.

The iron rule demands that scientific arguments consider only the explanatory power of contending theories. The positive core of the rule is a shallow, permissive conception of explanatory power, on which a phenomenon is explained by deriving it from a theory's causal principles. The principles need not pass any philosophical test or even be fully understood—thus, Newton considered himself to have explained the motions of the planets and the tides using his theory of gravity, although he offered no explanation of the causes of gravity itself.

The negative side of the rule forbids scientists, when making their case in official venues such as scientific journals, to assess theories using anything other than explanatory power. Philosophical and religious arguments in particular are out of bounds, no matter how compelling they may seem to scientists and to society at large. Likewise, scientists may not bring personal or cultural or other parochial considerations to bear in making their case; the iron rule requires that everything subjective be removed from scientific argument.

The Scientific Revolution, then, accomplished by way of the iron rule both a shallowing and a narrowing of the old forms of deliberation: post-Revolutionary argument is shallower in its conception of explanatory power, and it is narrower in its range of reasons for accepting and rejecting hypotheses and theories. Although such constrictions have little intuitive appeal, they have turned out to provide the superstructure for an extraordinarily effective engine of inquiry.

We live in a Tychonic world—a world in which great competing stories about the underlying nature of things can be distinguished by, and only by, scrutinizing subtle intricacies and minute differences. Humans in their natural state are not much disposed to attend to such trifles. But they love to win. The procedural consensus imposed by the iron rule creates a dramatic contest within which the trifles acquire an unnatural luster, becoming, for their tactical worth, objects of fierce desire. The rule in this way redirects great quantities of energy that might have gone toward philosophical or other forms of argument into empirical testing. Modern science's human raw material is molded into a strike force of unnervingly single-minded observers, measurers, and experimenters, generating a vast, detailed, varied, discriminating stock of evidence.

At the same time, the iron rule preserves this evidence, maintaining a craft tradition of "sterilization" that archives observed phenomena in a form that is distorted as little as possible by interpretation and other consequences of plausibility rankings.

The thinking of each generation of scientists is, and is permitted by the iron rule to be, essentially subjective. But that subjectivity does not matter in the long run. As thinkers come and go, observations accrue, revealing in time which theories are better explainers and which are worse. The eventual consequence is Baconian convergence on the truth: informed opinion increasingly favors the one theory, the correct theory, that accounts for every aspect of the accumulated evidence.

Science, then, is built up like a coral reef. Individual scientists are the polyps, secreting a shelly carapace that they bequeath to the reef upon their departure. That carapace is the sterilized public record of their research, a compilation of observation or experimentation and the explanatory derivation, where possible, of the data from known theories and auxiliary assumptions. The scientist, like a polyp, is a complete living thing, all too human in just the ways that the historians and sociologists of science have described. When the organism goes, however, its humanity goes with it. What is left is the evidential exoskeleton of a scientific career. You can see the bare bones laid down by Eddington's eclipse expedition, for example, in the black-and-white rows of numbers

that represent stars' photographed positions and the mathematical calculations that yield the sun's implied bending of starlight (Figures 2.3
and 7.4).

The intellectual edifice that is scientific knowledge is composed
largely of these exoskeletal remains. It is held together, like a reef, not
by living things, but by the evidence and argument that living things
produce, assembled according to a strict architectural plan ordained by
the iron rule.

Look at science's theater of inquiry, and you see life. You see the
surface of the reef, where the polyps still thrive: you see working scientists going about their investigations guided by hunches, intuitions,
ambitions, temperament, circumstance, culture. If you were to suppose
that this was science in its entirety, you would conclude it was subjective through and through. That is the characteristic mistake, I think, of
those radical subjectivists who infer from the day-to-day contextuality
of scientific activity a long-term epoch-to-epoch contextuality. Neither
the reef nor science is as lively, as soft, as transient as it appears. Both
are built deep down from simpler stuff that outlasts the passing of the
organic profusion on the periphery, forming a grand, stony, and severe
structure upon which rests their success.

III

WHY SCIENCE TOOK
SO LONG

CHAPTER 9

Science's Strategic Irrationality

ତ୍ତ୍

Why did science take so long to invent?
Because the iron rule looks like a terrible idea.

"CREDIT MUST BE GIVEN . . . to theories only if what they affirm agrees with the observed facts." That sounds as scientific as anything said by Isaac Newton or Richard Feynman. It was written by Aristotle in the fourth century BCE. Aristotle was a systematic observer, an innovative theorist, and a first-rate intellect. He put the highest priority on a theory's ability to explain the phenomena. What stopped him from inventing modern science two thousand years before the Scientific Revolution?

It is tempting to answer this question by writing a history of the Scientific Revolution, pointing to social, intellectual, and economic factors at work in the seventeenth century that were almost or entirely absent in Aristotle's era: the preeminence of the mechanistic philosophy propounded by Descartes and Boyle, the application of mathematics to physical theorizing, new technologies such as the printing press and improved lens-grinding and glassblowing techniques, an emerging regard for inquiry into nature as a high-minded form of religious worship.

It makes for a fascinating story. A narrative of this sort, however, is not very well able to distinguish what is essential to modern science from what was essential to modern science's developing in the particular way that it did in the seventeenth century. Certainly, there could have been no Galileo or Newton without mathematics, but not all scientific inquiry requires advanced math. (The development of chemistry did not.) There could have been no Boyle or Huygens without atomism, but not all science is about molecular interactions. (The theory of evolution by natural selection is not.) The telescope, in the hands of Galileo and others, rocked seventeenth-century astronomy, yet Tycho Brahe's naked-eye observations of the planets would have been sufficient on their own to sustain the new physics of gravity.

To determine why science took so long to arrive, then, I started not with a historical narrative but with a philosophical examination of modern science as a whole, seeking out its sine qua non: the thing or things that are decisive in explaining its powers of discovery. That sine qua non has turned out to be the iron rule of explanation. Aristotle and so many other natural philosophers, in many places and at many times, failed to set in motion the knowledge machine that we call modern science because, for all their concern with observation, they failed to invent the iron rule.

Compare Aristotle with Newton. Both were aiming for the same goal, a grand theory that explained the way things move and change. Their methods, however, were quite different. Aristotle but not Newton subjected his hypotheses to stringent philosophical tests. Newton but not Aristotle subjected his hypotheses to stringent quantitative tests, demanding they explain not only the qualities of motion—circular versus straight, up versus down—but also the finest details, such as the precise trajectories of the planets captured by Kepler's laws.

The quantitative tests, it turned out, were far more important. Although Aristotle carefully scrutinized all manner of natural phenomena and cared deeply about the power of his hypotheses to explain what he saw, he was typically content with accounting for broad patterns and not matters of particular detail. It is on those small facts, however, that much of science's truth-finding power hinges.

It is extraordinarily difficult, as Kuhn saw, for human beings to maintain any kind of prolonged interest in the all-important small facts. Minutiae are seldom intrinsically interesting; to transcribe them from nature is often tremendously hard work; and grand intellectual endeavors—conceptual, philosophical, systematizing—are always beckoning, luring the empirical investigator away from the lab bench or the observation post to the glittering land of ideas. Most scientific work is more like bookkeeping or long division than it is like poetic self-expression or polar exploration. Great minds are hardly likely to seek out such a life.

The iron rule solves this problem not by attempting to glamorize what's clearly menial, but through a more indirect, more devious stratagem. It sets up scientific argument, as I have explained, as a kind of game in which hypotheses are defended and attacked. In that game, only one kind of move is legitimate: the empirical move in which a hypothesis is attacked for failing to explain some observed matter of fact and defended by showing that the failure is merely apparent, due to malfunctioning equipment, unfavorable conditions, or faulty assumptions. Victory does not come through smooth rhetoric, metaphysical inquiry, moralizing, or any other sort of sweet talking or big thinking. To win, players must front up with meticulous observations.

For Aristotle to subject himself to the iron rule, he would have had to subsume his physical, chemical, biological, psychological, and astronomical investigations under the regulations of this game. Like every avid player, he would have had to put aside the search for higher harmony. The physical and intellectual energy thereby liberated would flow in the only remaining direction, toward unrelenting empirical testing. By the time of his death in 322 BCE, a classical Scientific Revolution might have been in full swing.

But Aristotle was a serious thinker; had such a game occurred to him, he would have dismissed it out of hand. He was not about to renounce philosophical reasoning.

He was no ideologue; he did not advocate philosophical reasoning as opposed to observation. Rather, he advocated philosophy *together with* observation—and when observation and philosophy clashed, Aristotle gave observation the upper hand. Criticizing some of his predecessors, he wrote:

> Their explanation of the phenomena is not consistent with the phenomena. And the reason is that their ultimate principles are wrongly assumed: they had certain predetermined views, and were resolved to bring everything into line with them. . . . As though some principles did not require to be judged from their results, and particularly from their final issue! And that issue . . . in the knowledge of nature is the phenomena . . . given by perception.

The iron rule would not, then, have offered Aristotle any new sources of information or any new ways to treat it; it would merely have told him to forget the philosophy. That looks like a rotten deal: dispensing with a valuable source of information, philosophical reasoning, without offering anything in return. It would have seemed unreasonable, barbaric, simply irrational.

The same is true of the iron rule's deliberate deafness to religious, spiritual, theological reasons for belief. Throughout history, the faithful have taken their religious doctrines to have implications for the material as well as the spiritual world and certainly for the domain of living things. The implicit clash with the iron rule is well dramatized by the intellectual agonies of William Whewell, Master of Cambridge's Trinity College from 1841 to 1866. Surveying the newly discovered fossil record, Whewell saw episodes of large-scale speciation that he thought could be explained only by God's intervention. A full understanding of the history of life would require, then, a unified treatment of geology, biology, and theology. Yet the iron rule, implanted in Whewell's mind from his undergraduate days, forbade him from providing such a synthesis in his scientific masterwork. It seemed wrong, even nonsensical, but he overrode his qualms, surrendering to the scientific method.

Take the roots of Whewell's quandary and project them back over the thousand years of history that preceded the development of modern science. Throughout the European Middle Ages and the roughly coterminous Islamic Golden Age, you find thinkers exploring the workings of the natural world by way of astronomy, optics, medicine, and more. None of these thinkers hit upon anything like the iron rule. They could not have. They were all, each in their own way, devout. It was clear to them that knowledge of God or of God's plan potentially had something to tell us about the way things were laid out in the physical world—in histories, living bodies, the system of planets. Perhaps theology would turn out to offer relatively little help to empirical inquiry, but the possibility ought to be explored.

The iron rule, however, firmly prohibits such exploration. In science, only empirical reasoning counts. Thus, to thinkers who unlike Whewell had not witnessed the rule supervise a string of stunning discoveries, scientific inquiry would have seemed a deliberately hobbled way of figuring out the world. An indiscriminate ban on theological thinking would have made no sense to these investigators, just as an indiscriminate ban on philosophical thinking would have made no sense to Aristotle. It is hardly surprising, then, that nothing like the iron rule arose in all the centuries spanned by Augustine, Avicenna, Averroes, and Aquinas.

Not even an early modern thinker like René Descartes, living on the cusp of the Scientific Revolution, would have been able to bear the iron rule's separation of empirical and theological inquiry. God's power and oversight is woven into Descartes's natural philosophy in numerous ways; it is God, for example, who ensures that our minds are stocked with concepts well suited to thinking about his creation, and it is God who sets the matter constituting that creation into uniform circular motion. From Descartes's point of view, it would have been quite unfathomable to forbid, as does the iron rule, taking into account the supremely important fact of God's existence and attributes—omniscience, omnipotence, benevolence—when investigating the natural world. And so we find in Descartes's writing, as we find in the writing of almost all those who came before him, not the slightest inclination toward the rule's absolutist decrees. In the 1640s, despite the best efforts of Bacon and a few other radicals, science was still reluctant to be invented.

This is why it arrived so late in human history: it seemed a cockeyed pursuit, an exercise in deliberate intellectual impoverishment. There was nothing wrong with the iron rule's emphasizing that observation matters, but everything wrong with its insisting that *only* observation matters—an anti-intellectual injunction to shut down every part of the head besides the eyes.

A vicious circle, then, cut off human minds from the scientific sensibility. There was no way to grasp how useful the iron rule would be without testing it in practice, but no reason to test it without first having some inkling how useful it could be. Indeed, there was good reason *not* to test it: it ignored what were regarded as indispensable sources of knowledge. The logic of the circle entrapped the ancient Greeks; it entrapped the medieval and early modern philosophers, such as Descartes. It equally entrapped, I conjecture, thinkers in China and Korea, in India and Persia, in Central America and the Andes. Had they considered the iron rule, they would have scoffed at it; almost certainly, so preposterous an idea never entered their minds.

IN 1859, 20 years after Whewell sought to interpret the fossil record using sophisticated creationist ideas, Charles Darwin published *On the Origin of Species.* Victorian science was electrified—and Victorian society scandalized—by the idea of evolution by natural selection. In the matter of creation, it seemed, God was being written out of the picture.

"The flood-gates of infidelity are open, and Atheism overwhelming is upon us," wrote the anguished evolutionist George Romanes in 1878. It was by no means all Darwin's doing. Geologists were casting doubt on the existence of a "Great Flood"; textual scholars were showing that the Bible was a compilation of texts written by many different authors at different times; French revolutionaries, English poets, and German social thinkers were imagining a world free of the strictures of organized religion. As the European nineteenth century progressed, the mantle of faith began to slip from the shoulders of the human spirit.

Today perhaps a third of American scientists believe in God. Even these believers do not for the most part think that their spiritual commitments could in any substantial way inform their research. They go along, for all practical purposes, with Stephen Jay Gould's contention that science and religion are legitimate forms of inquiry into two entirely distinct subject matters, that they are "non-overlapping magisteria." If you want to understand the meaning of life, by all means take up religion. To understand the movements of the planets and the origin of species, however, empirical observation is all you need. For the great majority of contemporary scientists, then, there is nothing in the least unreasonable about the iron rule's exclusion of religious considerations from scientific argument.

The same is true of the rule's exclusion of philosophical argument. Most physicists regard it as a waste of time, for example, to search for an understanding of quantum mechanics that renders it humanly comprehensible. Just use its mathematical machinery to make predictions and to construct shallow explanations, they say—"Shut up and calculate." The physicist Steven Weinberg goes further:

I know of *no one* who has participated actively in the advance of physics in the postwar period whose research has been significantly helped by the work of philosophers.

Does the perceived irrelevance of religious and philosophical considerations in contemporary science mean that the iron rule has lost its semblance of logical perversity? That it no longer appears to censor sensible nonobservational reasons for belief? Then we would have a science admirable not only for its abundant successes but also as a paragon of rationality.

Yet we do not: even today the iron rule has an air of unreason. I'm not thinking of the handful of scientists who see their God working manifestly in the material world or who seek to philosophize their way to theoretical understanding. I have in mind rather a form of reasoning that a great number of contemporary scientists, including Weinberg himself, celebrate for its power to illuminate the natural world: arguing for a theory's truth on the grounds of its beauty.

The iron rule proscribes the use of all such arguments in science's official channels of communication. Consequently, any modern investigator who believes that aesthetic qualities can show us the way to truth must consider the scientific method to be willfully ignoring something of significant value. In that case the method would be, in spite of its productivity, strictly speaking in violation of the principles of reason. In the coming chapter I will examine the iron rule's war against beauty and the implications for its rationality more closely. Is it the rule that's irrational? Or those who believe in the guiding power of beauty?

CHAPTER 10

The War against Beauty

ᘒᘓ

Appeals to aesthetics have no place in
public scientific argument, insists the iron rule.
This ban on beauty is also an attack on reason:
elegance often, if not always, points the way to truth.

BALANCE, BEAUTY, SYMMETRY. For all its dust, dirt, and motley ways of death, the universe has a good measure of these ethereal properties—a nobility of structure and an elegance in the fine texture, or so most of us would like to believe. There is a way in which, considered as an aesthetic whole, the world makes sense.

This hidden harmony is a key to a deeper understanding of things, many great thinkers both before and after the Scientific Revolution have supposed. Find the patterns governing nature's inner principles, and you can call on your sense of proportion, your eye for grace and elegance, to help discern the composition and the causal underpinnings of the universe. A good theory must explain the observable facts, but alongside explanatory power, beauty, too, shows the way to truth.

And yet, like philosophical and theological reasoning, aesthetic reasoning is eliminated entirely and indiscriminately from official scientific argument, from the journals and conference presentations, by the iron rule. It is a highly functional prohibition, an important part of the program of manipulation by which science molds ordinary human minds into indefatigable empirical-testing machines. But is it reasonable? To answer that question, we need to know just how effective a guide to truth beauty has turned out to be. We should therefore consult the historical record.

THE CONSONANCE OF microcosm and macrocosm—that is the most ancient principle of universal harmony, and it was for a long time the most commanding. The human world reflects the physical world, proclaims this principle; further, the same patterns marshal the connections between things at every level of being: physical, chemical, biological, psychological, and theological; on earth as it is in heaven, among angels as it is among men.

A taste of this view's strange charm—a single pattern imposing itself at every level—still lingers today in the fascination with fractals. Many examples of these geometrical figures contain smaller embedded images of themselves, which then contain smaller embedded images still, and so on ad infinitum. Look into a well-known fractal, the triangle described by the Polish mathematician Waclaw Sierpinski in 1915 (Figure 10.1). You will see that within the outer triangular boundary, each of the three next largest triangles is an exact copy of the whole. Thus, the same triangular structure repeats itself at every scale, high and low, macro and micro. There is no smallest triangle: an infinity of ever more diminutive shapes is generated by simple three-sided iteration and miniaturization. It is a most elegant way to make a universe. Once upon a time, this was considered not merely a mathematical curiosity but the secret to all creation.

What is the master plan? One answer is provided by the venerable rule of four, a principle dating from antiquity that finds fourhood throughout God's creation. The world is made of four elements: fire, air, earth, and water. The human body is controlled by four vital flu-

Figure 10.1. The Sierpinski triangle.

ids: blood, black bile, yellow bile, and phlegm. There are four cardinal virtues (justice, moderation, fortitude, and prudence), four evangelists, four letters in the name of god (in Latin, *Deus*), four seasons, four principal winds, four regions of the world, and four ages of man (childhood, adolescence, adulthood, old age). The same tetradic configuration is replicated at every level, from the grand structure of the cosmos through meteorological dynamics down to the physical and moral makeup of human beings and their spiritual history and destiny.

An even more fundamental fourhood underlies all of this. There are four physical attributes—hot and cold, moist and dry—that are found mixed in each of the substantive quartets. Fire is hot and dry, as are summer and adolescence; water is cold and moist, as are winter and old age.

This vision of universal fourfold harmony is dramatically expressed in the diagram with which the medieval English monk Byrhtferth (c. 970–c. 1020) concluded his treatise on the determination of the date of Easter (Figure 10.2). Looking deep into the diagram, you can unravel further implications of the scheme of four; for example, the initial letters of the four regions (Anathole, Disis, Arcton, Mesembrios), as Byrhtferth points out, spell Adam, the name of the first human and, for better and worse, the first inquirer after knowledge (or, more exactly, her assiduous assistant).

Similar organizing structures are strewn like spiky bouquets across the history of human thought, projecting their internal symmetry onto the sky above and the world around. The Swiss Renaissance thinker Paracelsus (1493–1541) saw a concordance of heavenly bodies with the parts of the human body: the sun, he thought, corresponds to the heart and to gold; the moon to the brain and to silver; the planet Venus to the kidneys and to copper; and so on for the other four known planets.

Isaac Newton eagerly consumed all the arcana of this sort that he could bring to light, his imagination seized by the idea of the world, its chemistry, its history, and its mythmaking as realizations of a great cosmic plan (Figure 10.3). The economist John Maynard Keynes, who purchased many of Newton's alchemical papers at a 1936 auction, vividly captured this aspect of Newton's worldview (alluding to Newton's concern with the *prisca sapientia*, the ancient wisdom):

> He looked on the whole universe and all that is in it *as a riddle*, as a secret which could be read by applying pure thought to certain evidence, certain mystic clues which God had laid about the world to allow a sort of philosopher's treasure hunt to the esoteric brotherhood. He believed that these clues were to be found partly in the evidence of the heavens and in the constitution of elements . . . but also partly in certain papers and traditions handed down by the

Figure 10.2. Byrhtferth's diagram. The four elements are *aer, ignis, terra,* and *aqua* (air, fire, earth, and water). The four seasons, which can be found in the circles between the elements, are *ver, estas, autumnus,* and *hiems* (spring, summer, fall, and winter). The four physical attributes (look along the sides of the diamonds) are *calidus, siccus, frigidus,* and *humidus* (hot, dry, cold, and moist).

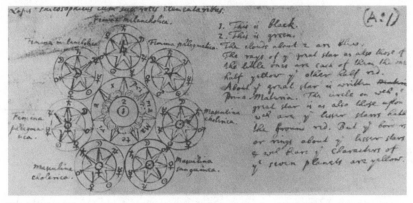

Figure 10.3. Newton's sketch (copied from an alchemical text) of the secrets of the philosophers' stone.

brethren in an unbroken chain back to the original cryptic revelation in Babylonia. He regarded the universe as a cryptogram set by the Almighty.

It is a kind of magical thinking, a kind of religious thinking, a kind of mystical thinking, but it is also a kind of aesthetic thinking, using the sense of beauty and structure to decode the cosmos.

It is not, however, an effective kind of thinking. Byrhtferth's numerology is delightful, but not insightful. That the iron rule excludes such methods from scientific argument cannot, given what we have seen so far, be held against it.

This is, however, only a single example of aesthetic evaluation put to work as a technique of inquiry. Before generalizing about the merits of such methods, we must survey a few more cases.

IN 1821, William Sharp Macleay, a British government official and a keen entomologist, published a study of insects in which he put forward a new framework for classifying life: the quinarian system. It took its name from the Latin word for the number five, and in that number lay its structural essence. Every biological taxon—the birds, the beetles, the big cats—Macleay believed, consisted of five subgroups, and each of those subgroups of five more subgroups, and so on down to the level of individual species.

By arranging any five sister groups in a circle, as shown in the classification of birds in Figure 10.4, further structure became evident.

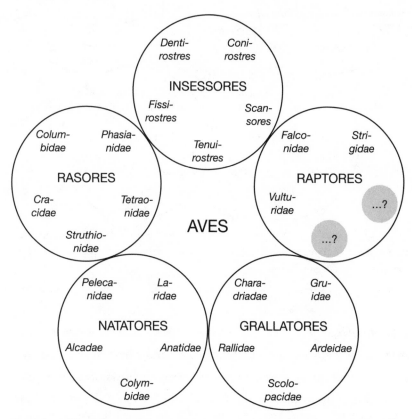

Figure 10.4. The top level of a quinarian classification of birds (Aves) by Nicholas Aylward Vigors. Affinities are found between neighbors within a circle (for example, Conirostres and Scansores in the top circle) and where distinct circles touch (for example, Scansores and Falconidae). In the circle of the Raptores only three positions are filled; this indicated the existence of two undiscovered taxa with appropriate affinities to the Strigidae, Vulturidae, and Gruidae (that is, owls, vultures, and cranes).

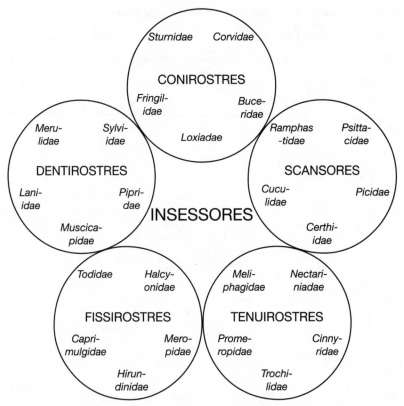

Figure 10.5. Vigors's quinarian classification of the Insessores, the group that appears at the top of the diagram in Figure 10.4. This figure shows the finer, deeper structure of the top circle in that diagram.

A group shared "affinities" with its immediate neighbors inside its circle, making for many important resemblances, and where the circles themselves touched, there were affinities between members of the groups on either side of the meeting point. In cases where a group appeared to contain fewer than five subgroups, such as the Raptores at the upper right of Figure 10.4, quinarians made a bold prediction: new taxa with the corresponding affinities would eventually be found.

The iteration of the quinarian principle creates a fractal-like geometry. Examine, for example, the diagram in Figure 10.5, which shows

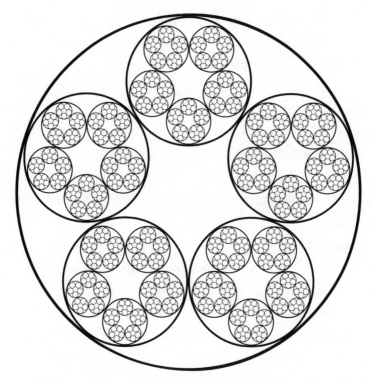

Figure 10.6. The nested structure of the quinarian system, shown schematically. Compare with Figure 10.1.

the Insessores from the top of Figure 10.4 in more detail. Not only do you see the five subgroups shown in the larger-scale figure; you see that each of these subgroups is itself divided into five more subgroups. As a consequence of this five-way nesting, life has the grand design shown in Figure 10.6, reminiscent of Figure 10.1's Sierpinski triangle.

The quinarian geometry and the affinities of neighboring groups were explained by Macleay as the consequence of a kind of branching process that generates life:

Nature appeared to me to have branched out in the animal kingdom . . . in a most beautiful and regular though intricate manner, that might be compared to those zoophytes which ramify in every direction, but of which the extreme fibers form by their connexion the most delicate circular reticulations.

Zoophytes are plantlike animals, such as corals or sea lilies; Macleay was imagining the quinarian pattern, then, as a cross section through a treelike growth that over and over branched five ways. He did not think of the branching as a historical event; it was rather an abstract mathematical template to which the living world was compelled to conform.

What discloses the existence of Macleay's supposed template, and so reveals a fundamental principle governing the biological world, is the human sense of beauty, drawn ineluctably to the fivefold symmetry that the quinarians discerned in the panoply of life. Is the iron rule, by ignoring the judgments of this aesthetic sense, thereby crippling our powers of discovery?

For a time Macleay's quinarian system was immensely popular in Britain. Many naturalists in the late 1820s and early 1830s looked for the characteristic iterated five-way structure in their own taxonomical bailiwicks. One such follower was the young Charles Darwin, who wrote over 80 pages in his "transmutation notebooks" attempting to reconcile quinarianism with his notion of evolutionary change. The branching structure effected by Darwin's "descent with modification" would naturally give rise to something with the nested, circular aspect of Figure 10.6, as Macleay had observed, but Darwin was stumped by the rigid law of five. Why would branching always take place in five directions, rather than, say, four or six? He made a brave attempt to extract fivehood from the process of environmental adaptation, but eventually gave up, writing in his notes, "Number five in each group absurd."

By then it was the spring of 1838, and quinarianism was fast falling out of favor. A more careful examination of many quinarian efforts
at classification showed that the five-way structure was often forced by
promoting or demoting groups to levels at which they did not genuinely belong, along with other forms of Procrustean taxonomical violence. Denouncing quinarianism, the influential naturalist Hugh Edwin
Strickland urged his colleagues

> to study Nature simply as she exists—to follow her through the
> wild luxuriance of her ramifications, instead of pruning and dis-
> torting the tree of organic affinities into the formal symmetry of a
> clipped yew-tree.

After Darwin's *On the Origin of Species* was published in 1859,
the reason for this "wild luxuriance" became clear: new kinds of
organisms evolve not to realize some universal mathematical imperative but to take advantage of local environmental opportunities or
to evade local environmental threats. William Whewell's friend, the
scientist John Herschel, somewhat derisively called natural selection the "law of higgledy-piggledy." He was right: natural selection's
tinkering, its jury-rigging in the face of immediate circumstances,
achieves marvelous transformations yet builds what is from the geometrical point of view a rather disorderly, if beguilingly organic, tree
of life (Figure 10.7).

The quinarian system was, like Byrhtferth's rule of four, a metaphysical fantasy. It might look also to be a fable with a methodological
moral: *Do not be seduced by formal beauty; in natural science, be guided only
by observed fact.* Another victory, then, for the rationality of the iron rule
and its spurning of aesthetics? For the value of strictly empirical thinking, for eyes trained on the raw data just as it lies and nothing beyond

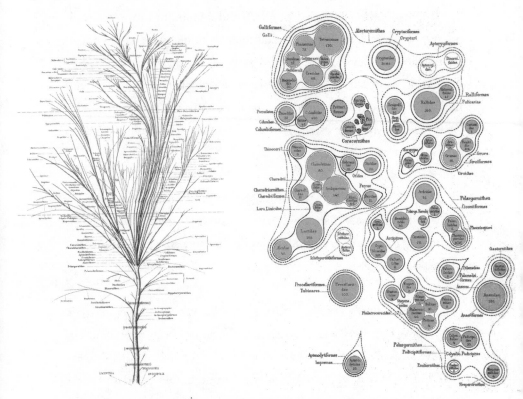

Figure 10.7. The tree of life after Darwin. Max Fürbringer's 1888 classification of birds shows none of the fixed five-way branching characteristic of the quinarian system. On the left, a view of the entire tree; on the right, a cross section whose irregularity can be compared with the symmetry of Figure 10.6.

or above? It is still too soon to reach a conclusion; we have seen only one side of the story. As some tales of success will reveal, there is in fact much to be gained in the study of nature if you keep your eyes open to the right sort of harmony and form.

THE SCOTTISH NATURALIST Sir D'Arcy Wentworth Thompson (1860–1948) laid down this credo:

> I know that in the study of material things, number, order and
> position are the threefold clue to exact knowledge; that these three,
> in the mathematician's hands, furnish the first outlines for a sketch
> of the Universe.

"Material things," for Thompson, included plants and animals, all of which reproduce and grow by way of physical mechanisms. Diverse biological forms are frequently generated by the same underlying causal process; in such cases there will be a hidden mathematical unity to be found in the diversity—so Thompson argued in his 1917 magnum opus *On Growth and Form.*

A study of the forms taken by soap bubbles and snowflakes, for example, revealed the physical basis of the multifarious and intricate skeletal structures of many kinds of radiolarians, single-celled organisms that make up a large proportion of oceanic plankton. Likewise, the same physical principles that determine the radial splash pattern made by a pebble falling into a glass of milk play a role, Thompson wrote, in the growth of the ring of tentacles that surround the mouths of polyps in the class Hydrozoa (some of which are the branching "zoophytes" that inspired the quinarian Macleay).

In his final and most famous chapter, Thompson introduced a technique of great elegance—an aesthetic power move, as it were—to uncover profound affinities between apparently distinct animal body types. The physical form of the fish *Polyprion* can be subjected to simple geometrical transformations, he showed, that as if by some mathematical magic turn it into each of three related species (Figure 10.8).

In an even more spectacular demonstration, Thompson used a simple stretching operation to morph the ordinary looking puffer fish into the biologically related but physiologically quite dissimilar and rather exotic sunfish (Figure 10.9). Of this latter case Thompson wrote:

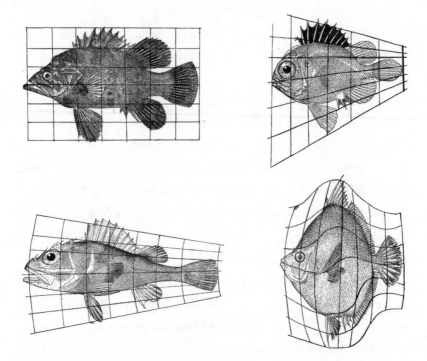

Figure 10.8. *Polyprion* (top left) transformed into three other kinds of fish. The transformations are simple stretchings of the square grid superimposed on *Polyprion*; imagine *Polyprion* drawn on a rubber sheet that is then deformed by being pulled in one or more different places and directions.

> It accounts, by one single integral transformation, . . . for the new and striking contour in all its essential details, of rounded body, exaggerated dorsal and ventral fins, and truncated tail.

Thompson proposed that the fish transformations, in their simplicity and beauty, suggested deep truths about development and growth, pointing to a single underlying physical mechanism at work in the related species—not Darwin's natural selection, but a mechanism that is prior to and that can be exploited by natural selection. Indeed, the transformations constituted a "proof" that a comprehensive "law of growth" has pervaded the whole structure in its integrity, and that some more or less simple and recognizable system of forces has been in control.

The same was true, Thompson believed, of the mathematically related physical forms that he found within groups of crustaceans, crocodiles, ungulates, and many other animals.

In contrast with his ideas about radiolarians and hydrozoans, he did not suggest any particular physical mechanism as the foundation of these laws of growth. His views did, however, enable him to make specific predictions. He correctly forecast that the human precursors known at the time—such as *Homo erectus* and the Neanderthals— would not form a "straight line of descent," but would rather be found on a complex tree of which humans were just a single branch (albeit the only branch still existing). He determined the likely form of transitional fossils linking one of the first birds, *Archaeopteryx*, with a later descendant, and he was able to use his mathematical methods to pre-

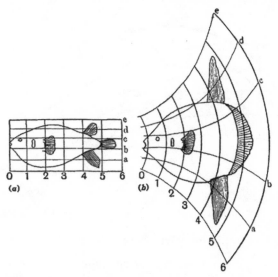

Figure 10.9. The puffer fish (left) transformed into a sunfish (right).

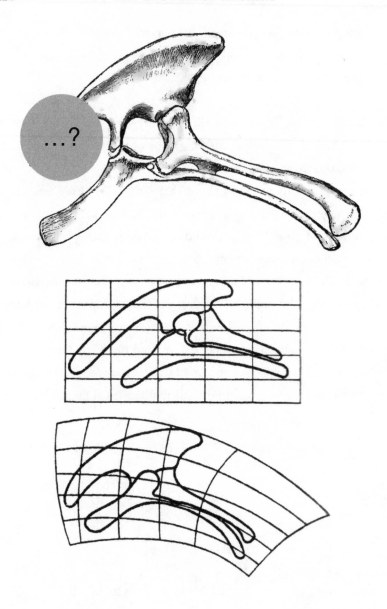

dict the shape of a missing part of the dinosaur *Camptosaurus*'s ilium bone (the upper pelvis) by extrapolating, using one of his mathematical transformations, from the known form of the dinosaur *Stegosaurus*'s ilium (Figure 10.10).

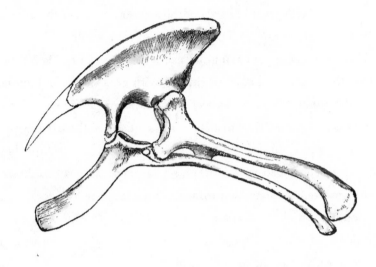

Figure 10.10. The reconstruction of the *Camptosaurus dispar* ilium bone. *Top left*: the fossilized ilium bone analyzed by O. C. Marsh, missing its front part. *Top right*: Marsh's reconstruction, in which the bone receives a sharp, toothy completion. *Bottom left*: Thompson's reconstruction (lower grid), based on a mathematical transformation of the same bone in *Stegosaurus* (upper grid). *Bottom right*: a more complete *Camptosaurus dispar* ilium, showing the superiority of Thompson's reconstruction to Marsh's.

We now know that the laws of growth are not, as Thompson supposed, simple physical principles, but are rather consequences of the convoluted mechanisms of developmental genetics. Even within this modern framework, however, Thompson's transformations are considered by many biologists to be part of the solution to one of the greatest explanatory puzzles raised by the evolution of complex life.

How does a body plan as unusual as that of the sunfish (Figure 10.9) evolve? This much we know. Beginning with an ancestor looking not unlike the puffer fish, natural selection must have adjusted many large-scale traits at the same time to achieve the sunfish's distinctive physique (and size—mature sunfish can weigh up to 2½ tons). If each of those traits were determined by a separate set of genes, such a coordinated adjustment would be extremely difficult, if not impossible: since natural selection acts on chance variations, it would have to wait for a happy coincidence of many independent mutations to produce even a single step along the way to the final product.

Thompson's discoveries suggest a more satisfying story, according to which strikingly different physiological layouts, such as the body plans of the puffer fish and the sunfish, are the consequence of small genetic differences that, in effect, control the knobs on the control panel of growth, imposing a "fun house mirror" effect on the finished product. Just one or a few mutations, then, might be sufficient to turn the dial ever so slightly toward the "sunfish" setting. A series of such events, in an environment that favors the sunfish way of life, provides a path to the evolution of these huge, strange creatures that is not so difficult for natural selection to follow.

More generally, the capacity to effect radical physiological change through minor genetic tweaks enables natural selection to easily explore numerous alternatives to a species' standard body plan, alternatives that are often weird yet occasionally highly advantageous. Had the capacity not existed, many evolutionary developmental biologists believe, complex life-forms could not have evolved. The earth would have had to rest content with its radiolarians and its polyps or perhaps something simpler still.

Thompson's insights were built, as he was proud to acknowledge, on his high regard for beauty and form. Cultivated in the right way and in the right place, then, a concern for mathematical simplicity and elegance is after all quite capable of illuminating the higgledy-piggledy world of biology.

Score one for beauty. To tip the argument decisively toward the truth-finding power of the aesthetic sense, however—and therefore to firmly establish the irrationality of the iron rule's scorn for beauty, the logical ugliness that made modern science so difficult to embrace—we should look to physics.

"IT IS MORE IMPORTANT to have beauty in one's equations than to have them fit experiment," wrote the English theoretical physicist Paul Dirac—the beauty being a sign that the theory was on the right track and that the discrepancy with experiment was likely "due to minor features . . . that will get cleared up with further developments." Beauty is the beacon; truth is what it marks. Einstein, according to the physicist Eugene Wigner, thought along the same lines: "The only physical theories which we are willing to accept are the beautiful ones." You needn't look far to find similar sentiments behind many doors in the corridors of theoretical physics or in the popular writing of physicists such as Subrahmanyan Chandrasekhar, David Deutsch, and Frank Wilczek.

The historical record suggests that they are onto something important. Newton's sense of beauty, it is clear, not only drew him to the mysteries of the *prisca sapientia*, the ancient knowledge, but also pointed him to the simple mathematics that lay behind the curves traced by cannonballs, planets, and comets, thereby unveiling the secrets of universal gravitation. Before him, Copernicus and Galileo were among science's beautiful thinkers, and after him, many more were to come. Indeed, there are too many cases of successful aesthetic reasoning in the annals of the physical sciences to list even in passing, so I will confine myself to a single momentous episode from the recent history of particle physics.

That will cement the case in favor of beauty and therefore against the rationality of the iron rule.

The story begins in 1931, when James Chadwick confirmed the existence of the neutron. With that discovery, the nature of the basic building blocks of matter seemed to have been settled. Atoms were made of central clusters of protons and neutrons surrounded by electrons, and electromagnetic radiation was made of photons, adding up to four fundamental particles in all. Byrhtferth the monk would have been gratified to see the rule of four making a smooth transition from ancient and medieval metaphysics to twentieth-century physics.

Or not entirely smooth: there were two additional particles that had been hypothesized but not yet seen: the pion, whose busywork was supposed to help keep the atomic nucleus together, and the neutrino, a small and mysterious particle emitted during a certain kind of radioactive decay. Still, even Aristotle had added a fifth element—the quintessence, of which he supposed the heavenly bodies were made—so why not six?

Had only the count stopped there. In the 1930s and 1940s, particle physicists headed for the alpine peaks of the Americas and Europe. Their aim was to get as close as possible to high-energy radiation pummeling the earth's upper atmosphere from unknown sources in outer space, so-called cosmic rays. There a bewildering spectacle lay in wait. When cosmic rays collided with air molecules, they created things that no one had ever seen before: entirely new particles, such as the muon, the kaon, and the mysterious lambda.

Then, with the development of sophisticated particle accelerators, such as the Cosmotron at Brookhaven National Laboratory, 60 miles east of New York City, the mountains became superfluous: weird particles could be generated in the outer suburbs of Long Island. Meanwhile, ever more sensitive detectors—notably, the bubble chamber invented in 1952 by Donald Glaser—made them easier to see. And new kinds kept appearing: the xi, the sigma, the delta, new kinds of pion and kaon, the eta meson. . . .

The variety was as dazzling as it was disconcerting, more reminiscent of the exuberance of a tropical rainforest than of the supposed bedrock of the universe. It was as though the biological law of higgledy-piggledy went all the way down—making for, as Robert Oppenheimer put it, a veritable "zoo" of particles. Would a bill of lading for the fundamental units of reality turn out to look more like a modern biological taxonomy—like the tree of birds shown in Figure 10.7—than like Byrhtferth's realization of the power of four? Or would someone find hidden order and simplicity in the zoo?

Murray Gell-Mann grew up in straitened circumstances in New York City, the child of immigrants from Eastern Europe who never quite secured their hold on the American dream. He was early on declared to be a prodigy, skipping grade after grade at school and entering Yale as an undergraduate at the age of 15. It was 1944, and much of the Yale campus had been co-opted for military training. Archaeology and linguistics were Gell-Mann's principal interests at the time, but his difficult and demanding father insisted he study something practical—if not engineering, then at least physics. As everyone around him was absorbed into the armed forces, the 15-year-old immersed himself in the subtle harmonies of the natural world.

A constellation of far-reaching discoveries lay waiting: thanks to his eye for underlying beauty, Gell-Mann would become one of the most important physicists of the twentieth century. He earned his doctoral degree in 1951, even as the gates of the particle zoo were swinging open and subjecting physicists to a stampede of new and peculiar forms of matter. Soon after, he conceived of a novel property of matter that he called "strangeness." Just as a particle can have a certain electric charge, it can have a certain quantity of strangeness, positive or negative. Protons and neutrons have no strangeness at all—or more exactly, they have a "strangeness number" of 0. But some of the new particles, beginning with the lambda, were truly strange. (The "strangeness number" of the lambda particle is -1.) When particles collide or decay, turning into other particles, strangeness is "approximately conserved": the sum of their strangeness numbers has a strong tendency to remain the same, before and after, as though any collection of matter contains a fixed quantity of strangeness that can easily be moved around but increased or decreased only with the greatest effort.

The postulation of strangeness brought some discipline to the zoo in the shape of a formula linking particles' strangeness to their electric charge and some of their other known properties. It also explained the mysteriously long time that it took the lambda particle to decay: it turned out that the decay can't happen without changing the amount of strangeness in the world. That doesn't prevent it, but it does throw on the brakes, allowing the lambda to linger for a few extra fractions of a nanosecond before disintegration.

Strangeness was just the first of Gell-Mann's insights into the structure of matter. In 1961, he published a geometrical organization of the hadrons, a class of particles that includes protons and neutrons and most of the other particles in the zoo. This scheme arranged the hadrons into several different groups according to what Gell-Mann called the "eightfold way."

You can think of the eightfold way as organizing a given set of eight particles—an "octet"—into a table, with rows corresponding to the particles' strangeness number and columns corresponding to their electric charge (Figure 10.11). There is a profound symmetry to octets that cannot be captured in such a figure—for that you need many more than two spatial dimensions—but which is mathematically encoded in what is called the SU(3) symmetry group. Suffice it to say that these arrangements of particles carried, in the eyes of Gell-Mann and other theoretical physicists, a kind of formal beauty, a mathematical rightness, that is fully the equal of the order on display in the quinarian system or in Byrhtferth's diagram.

Gell-Mann went on within a couple of years to reason that his octets, together with a "decuplet" of 10 more exotic particles, pointed to a deeper organizing principle in nature. In the same way that the complicated structure of Byrhtferth's diagram is generated by a quartet— hot, dry, moist, and cold—and the quinarians' structures are generated by an iterated fivefold symmetry, so the octets and decuplets were, to those with mathematical eyes to see their SU(3) symmetry, generated by

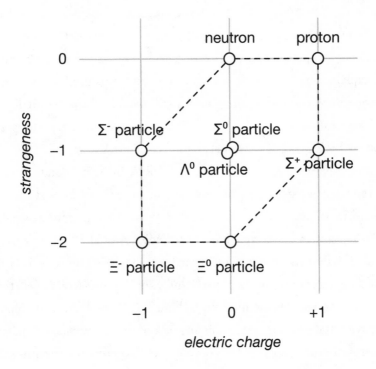

Figure 10.11. The Gell-Mann octet containing the proton and the neutron. Rows share the same value for strangeness; columns share the same value for charge. At the center is the lambda particle, labeled Λ^0.

a principle of three, a triplet called the "fundamental representation" of SU(3). Gell-Mann seized on this three-ness, hypothesizing that all the particles in his octets and decuplets were built from more fundamental particles, the quarks, of which there were three kinds: up, down, and strange (along with their "antiparticles"). A proton, for example, is made of two "up" quarks and a "down" quark, while a positive pion is made of one up quark and one down antiquark. The once sui generis lambda particle turns out to be nothing more than a coalition of an up quark, a down quark, and (of course) a strange quark.

The pursuit of beauty had led to truth: almost all contemporary scientists accept the reality of quarks (of which it now turns out there are six: Gell-Mann's three and then three more far larger ones). In 1961, however, Gell-Mann could not argue for the eightfold way on the basis of its charismatic symmetry, its gratification of the aesthetic sense. Hard evidence was needed. A part of that evidence was the theory's ability to explain the observed properties of the numerous particles known to belong to the zoo; far more glorious still would be a novel prediction subsequently vindicated by observation. From the empirical as opposed the aesthetic point of view, that would be the mark of truth.

In July 1962, at a conference at CERN, in Switzerland, the discoveries of two new xi particles were announced: the Ξ^{*-} and the Ξ^{*0}. Gell-Mann, sitting in the audience that day, realized that the new particles, along with 7 that were previously known, would fill all but a single space in a potential decuplet of 10 particles, shown in Figure 10.12. Like the quinarians positing additional families of Raptores to complete their elegant five-way taxonomic circles (Figure 10.4) or D'Arcy Thompson inferring the structure of the missing part of the *Camptosaurus* hip bone to preserve the simplicity and continuity of his geometrical transformations (Figure 10.10), Gell-Mann reasoned that this additional particle must exist in order to uphold the beauty of the eightfold way. During a discussion period, he sauntered to the chalkboard, drew the decuplet with its suggestive gap, and then filled the gap by boldly predicting the existence of a new particle with a certain charge, strangeness, and mass, the omega-minus—so-called because omega, the last letter of the Greek alphabet, signaled the structure's consummation.

The missing particle turned up on Long Island in 1964, thrust into existence by the Alternating Gradient Synchrotron and observed in a big new bubble chamber at the Brookhaven National Lab. It was exactly what the iron rule demanded: an empirical test. Gell-Mann had predicted the omega-minus, and sure enough it—or rather its distinctive bubble-chamber signature, shown in Figure 10.13—had been observed. Murray Gell-Mann had secured his Nobel Prize.

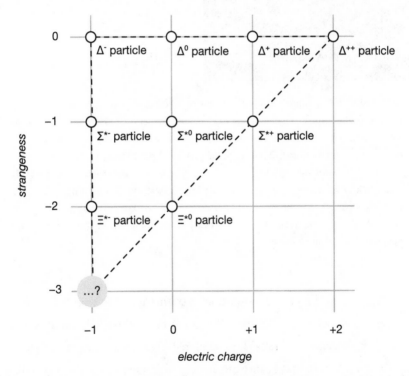

Figure 10.12. A decuplet that would be completed by a particle filling the gray zone at the bottom left, that is, a negatively charged particle with a strangeness number of –3. Gell-Mann predicted the existence of such a particle: the omega-minus.

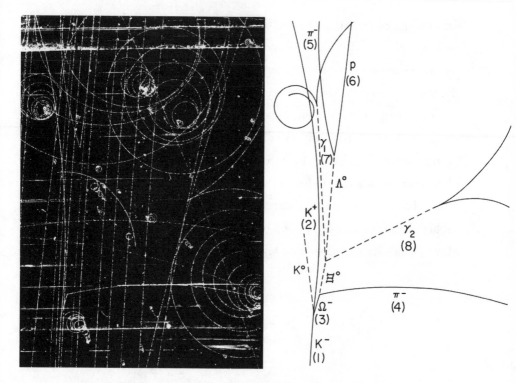

Figure 10.13. What counts in scientific argument: the telltale signature of the omega-minus particle. On the left is a photograph of the bubble-chamber tracks recording the trajectories of various particles created by the collision of a proton with a kaon. Only electrically charged particles show up in a bubble chamber; neutral particles are invisible and so their trajectories must be inferred. On the right is a schematic of the crucial elements of the image (with the inferred trajectories of neutral particles shown as dashed lines). The kaon (K^-, numbered 1 in the schematic) enters at the bottom. The omega-minus itself (Ω^-, numbered 3) exists only for a short time before decaying into an inferred xi particle (Ξ^0) and a negative pion (π^-, numbered 4).

I want to direct your attention past this glorious pageant of discovery to something rather peculiar going on in the background. On a number of occasions, Gell-Mann declared his allegiance to the Platonic precept that truth and beauty are entangled, saying in an informal talk, for example, that beauty, simplicity, and elegance are "a chief criterion for the selection of the correct hypothesis." Yet he made no appeal to this criterion in his official publications on the eightfold way. Nor do his physicist colleagues. Flip through the pages of the journals *Physical Review* or *Physics Letters*; you will find no invocations of beauty, no arguments grounded on a theory's grace or charm, though the authors might covertly hope that readers' tastes will resonate with their own. The reason is the iron rule's edict that only empirical testing counts.

The cosmologist Brian Greene sums up this consequence of the iron rule's tunnel vision succinctly in his book *The Elegant Universe*. Although physicists "make choices and exercise judgments about the research direction in which to take [a] partially completed theory" that are sometimes "founded upon an aesthetic sense—a sense of which theories have an elegance and beauty of structure on par with the world we experience," nevertheless,

> aesthetic judgments do not arbitrate scientific discourse. Ultimately, theories are judged by how they fare when faced with . . . hard experimental facts.

That is an acute characterization of an elegance-loving scientist's awkward existence under the iron rule: in your private thinking, let a theory's grace and beauty convince you of its truth, but say nothing whatsoever of these qualities when you go to persuade others. Like Whewell, deprived of his right to reason theologically in his scientific work, the aestheticizing scientist cannot publicly—or at least cannot officially—call on their most cherished springs of insight.

DOES IT MAKE SENSE? From a practical point of view, certainly: the injunction against aesthetic disputation forces all scientists, no matter how enamored with beauty, to put their energy if not their passion into the prediction and generation of empirical detail. But does scientific argument conducted according to such a precept respect the rules of logic? Not if it constitutes a categorical rejection of useful information. The question, then, is how valuable, on balance, aesthetic considerations tend to be.

To reckon the score, look at the history. On the one hand, the idea that the microcosm mirrors the macrocosm was a thesis that went nowhere. It was not a ludicrous notion: for all that anyone knew in 1600, God might have taken great pleasure in recapitulating his creation according to the same plan at every scale. Newton cannot be criticized for hedging his bets between experimentation and interpretation. But in the end, the beauty of fractals aside, there turned out to be meager profit in searching for profound parallels between the great and the small. The same is true for the Pythagorean approach of the quinarians. In the wake of Darwin, it became clear that there would be no universal formal structure in the organization of species, but only Herschel's law of higgledy-piggledy.

On the other hand, there are the successes: D'Arcy Thompson's mathematical transformations of body type, Gell-Mann's eightfold way and his system of quarks, and before them, a string of insights in mathematical physics, chemistry, and other sciences. The failures are on an epic scale, yet the successes are too important, too frequent, and too dazzling to ignore.

Scientific inquiry should therefore take an open-minded attitude toward aesthetic considerations. It should permit beauty in all its forms to provide guidance in concert with the observable facts. As more is learned, a scorebook should be maintained, ranking different styles of aesthetic thinking according to their successes and failures in the various domains of scientific inquiry. Striking symmetries, it might emerge, are tremendously important in fundamental physics, sometimes revelatory in physiology, but almost worthless in biological taxonomy. A fixation on simple arithmetic—as manifested in the ancient law of four depicted in Byrhtferth's diagram or the law of five underlying the quinarian system—might turn out to be useless everywhere. The creator of the universe, it seems, is not that sort of mathematician. With time, science could learn to distinguish the dangerously grandiose numerology of the quinarian system and its like from the subtle and fruitful attention to order and symmetry that brought us the theory of quarks.

The iron rule does not take this reasonable path. It is painfully simple-minded, uninterested in making sophisticated distinctions; it shuts down aesthetic deliberation altogether, rejecting the good along with the bad.

Suppose you are handed an envelope. You know that it is more than likely to contain information that is valuable, even crucial, to realizing your goals. Surely you open it? When the envelope is packed with aesthetic guidance, the iron rule throws it away. That is illogical, unreasonable, irrational.

A SCIENTIFIC RATIONALIST like Popper might regard the iron rule, with its exaltation of empirical testing, as a gleaming instrument of reason, paring away meddlesome theology, feckless philosophy, and a sentimental weakness for beauty—not painlessly, perhaps, but with an inexorable logic that no one could justly resist. I have said by contrast that the rule's ham-fisted surgery violates the principles of rationality. Some of what it removes we are well to be rid of, but it goes too far, and goes too far in an especially simplistic and thoughtless way: it imposes a wholesale prohibition on all forms of nonempirical thinking, no matter how judicious, no matter what their track record, no matter how well they synergize with empirical observation. The iron rule takes to human thought not with a logical scalpel but with a cleaver.

The progress enabled by this butchery cannot be denied. The iron rule may impose irrationality on scientific argument, but it is a strategically brilliant irrationality, manipulating its human subjects into the vehement engagement with empirical detail that makes modern science so formidable a knowledge-making machine.

The first time I read Kuhn's *Structure of Scientific Revolutions*, I paid too much attention to the titular revolutions and missed what in the book was truly revolutionary: the suggestion that science, humanity's supreme rational achievement, is driven by a kind of narrowness or blindness. It was during my second reading—on a train somewhere between Philadelphia and New York City, I still remember—that I looked out the coach window and saw the book in an entirely new way.

Kuhn was wrong about paradigms. Scientists' private thinking is not locked down, not limited, in the manner he supposes. What is severely limited is scientists' official speech—deprived of philosophy, religion, and beauty—and those limits, irrationally narrow in their own way, have the same ultimate effect that Kuhn attributed to the paradigm: to launch a fiercer, more tenacious, and more exacting search of the space of possible explanations than would have been attainable, given the foibles of our psychology, by any other means. That is, all in one, a fabulous paradox, a lesson in humility, and a reminder that humans are very interesting people.

An apparently impenetrable logical barrier once stood between those people and modern science. The iron rule, science's first commandment, looked perverse; perceiving its perversity, if they considered it at all, knowledge seekers had no reason to put it into practice; not seeing it at work, they could have no notion of its power. If the human race was going to get its vaccines, its electric motors, its wireless communicators—the wellsprings of health, the armatures of industry, the filaments of human connectivity—something out of the ordinary had to shatter the barrier. Newton's unusual psyche, his instinctive urge to compartmentalize, was the right kind of eccentricity. But he could not have broken through on his own; there is a larger story to tell about a sudden change of hearts and minds in seventeenth-century Western Europe.

CHAPTER 11

The Advent of Science

ℚ℧ℴ

Why, when science finally arrived, it was in
Western Europe and not some other place and in the
seventeenth century and not some other time

W RAP YOUR FINGERS AROUND an Acheulean hand ax, and
you are holding one of the oldest of all human technologies
(Figure 11.1). For over a million years, *Homo erectus* and its progeny used
these tools to butcher animals, scrape and slice their hides, cut wood, dig
roots, and much more.

Anthropologists from another star system visiting this planet not
so long ago might have supposed that the human lineage would never
change its ways. What had worked for a million years would work for
a million more: the Acheulean axes would be standard equipment for
intelligent life on earth forever.

Not so. Around 300,000 years ago, a dramatically superior stone
technology appeared in Europe—what are called Mousterian axes,
scrapers, and spearpoints (Figure 11.2). What was responsible for this
sudden innovation? Was there a Newton of stone, a reclusive, cave-
dwelling genius who saw the way to make the leap from the Acheulean
to the Mousterian way of hammering and flaking flint after intermina-
ble iterations of the same old thing?

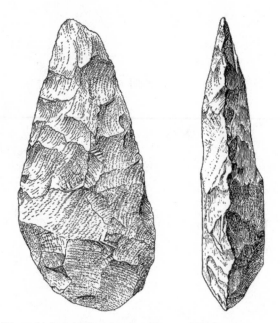

Figure 11.1. Acheulean hand ax (front and side).

The archaeological evidence suggests a rather different answer: Mousterian technology appeared with the evolution of a wholly new species—the Neanderthals. These low-slung, craggy-browed individuals were long treated as the archetypal cave people—mute, simpleminded, unreflective club bearers. The more we learn about them, however, the less accurate this picture turns out to be. The Neanderthals shared something like 99.7 percent of their DNA with modern humans, and they were human in many far more meaningful ways: they were not mute, but had language after all, along with art, ritual, and complex hunting techniques. The transition from *Homo erectus* to *Homo neanderthalensis* seems to have hinged on the evolution of a new kind of mind. It was the superior power of these enhanced organs of thought, perhaps, that enabled the invention of Mousterian tools. At any rate, after this great cognitive leap forward, Stone Age technology began to evolve at a newly

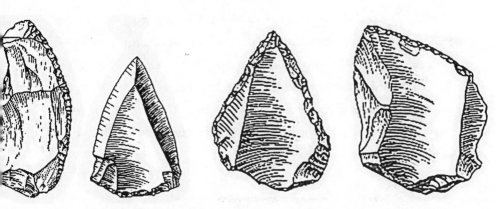

Figure 11.2. Mousterian tools: from left, a convex sidescraper, a Levallois spearpoint, a Mousterian spearpoint, a canted scraper.

rapid pace: gaps between innovations were measured not in millions of years but in tens of thousands of years, with tools becoming ever lighter, sharper, and more incisive.

Mousterian tools first appear in the archaeological record 300 millennia ago. Humanity's keenest, subtlest, most sophisticated tool of all—modern science—first appeared just over 300 years ago. Its belated debut poses the same question as the Mousterian explosion: why, after such prolonged stasis, so sudden a breakthrough?

Evolution, in the case of science, is certainly not the answer. Natural selection made no biological breakthroughs in the seventeenth century. However modern science was made, it was done with the same brains that the human race had been wielding for tens of thousands of years. Likewise, it was done with cultural tools that had been part of the human kit for many centuries before science: philosophy, logic and mathematics, systems of weights and measures, the rule of law and the division of labor. None of this, evidently, was sufficient in itself to set off the Scientific Revolution. Something else must have sparked the fire.

If *The Knowledge Machine*'s explanation of how modern science works is on the right track, then the decisive innovation of the Scientific

Revolution was the iron rule of explanation. The narrowing of discourse required by the rule amounted to a demand for flagrant irrationality—a demand that for a long time constituted an immovable mental block to humanity's willingness to take the iron rule to heart. There is, then, one question about the appearance of modern science that is more important than any other: how did the iron rule's irrationality become, all of a sudden, unobjectionable, even enchanting?

IN 1517, Martin Luther planted his "Ninety-Five Theses" on the door of a church in Wittenberg, Germany, precipitating a decisive break with the Catholic Church, the institution that had for one thousand years ruled religious life in Western Europe. In Italy, meanwhile, artists such as Piero della Francesca and Leonardo da Vinci developed the technology of perspective and depicted the human form in novel and more naturalistic ways, while across Europe scholars, builders, and writers rejected medieval models in favor of the literature and architecture of the classical world of Greece and Rome.

Perhaps the most radical of all these attempts to rebuild and rethink the highest forms of human activity were the efforts of the philosophers. To put human knowledge on a solid footing, René Descartes wrote, "I realized that it was necessary . . . to demolish everything completely and start again right from the foundations." With these words he exemplifies what is most exhilarating about Renaissance and early modern thought: the rejection of prevailing intellectual authority, above all the authority of the Catholic Church and its official philosopher, Aristotle, and a subsequent determination to construct everything from scratch and on its intrinsic merits, giving no hypothesis special consideration merely on the grounds of longevity, eminence, or religious orthodoxy. Not only Descartes, but all the great seventeenth-century philosophers—among them, Thomas Hobbes, G. W. Leibniz, and Baruch Spinoza—gave over their lives to the creation of new philosophical systems.

One of those systems rejected philosophy itself. By the seventeenth century, natural philosophers had been wrestling with the problem of the world's unobservable structure for two thousand years, with rather limited success. Figuring that it was time for something utterly different, a small group of thinkers placed a bold wager on a narrowly empirical form of inquiry. I am referring, of course, to writers such as Galileo Galilei, Robert Boyle, and Isaac Newton—and leading the charge Francis Bacon, who in 1620 so forcefully repudiated Aristotle's metaphysics in favor of the strategy of judging a theory entirely by its ability to explain the observed facts: "A new beginning has to be made from the most basic foundations."

This, then, is how the notion that "only empirical evidence counts" found its way into the heady seventeenth-century mix: not because the times especially favored empiricism and certainly not because they were hostile to philosophical or theological argumentation—nothing could be further from the truth—but because they were unusually fertile for bold thinking of every sort. The garden of the human intellect exploded in a profusion of ideas without precedent in human history. In the midst of such tremendous variety, something like the iron rule of explanation might be expected, if only for a season, to flower.

At this point, finding the principal elements of the iron rule spelled out in the writings of Bacon or Boyle, you might suppose that the advent of science is explained. In fact, the explanation has barely begun. The dominance of the iron rule, even once formulated, was hardly preordained; there was little reason to think that it would take over the garden. It is one thing to flower, quite another to flourish. Bacon laid out some of the principles of iron rule–governed science, but he did not put them to work in a concerted way. Galileo and Boyle did better, generating and explaining great amounts of quantitative data. Yet ideologically, they belong with Descartes as much as with Newton. Boyle wrote a great deal reporting his experimental investigations, but he wrote more still arguing like Descartes that all scientific explanation must proceed on the atomistic model, accounting for every observable fact in terms of "little bodies variously figured and moved." Rival styles of explanation, most of all the Aristotelian philosopher's appeal to "occult qualities," were, according to Boyle, "unintelligible." Boyle was not a systematic philosopher, but he used philosophical argument to promote what he considered to be a single legitimate explanatory style. We are still very far from Newton's explanatory permissiveness.

Deprive yourself of hindsight, and it is easy to imagine natural philosophy returning, by the late seventeenth century, to a status quo not so different from Aristotle's or Descartes's ideal, a theory of nature explaining the qualitative and some quantitative facts but always within a carefully reasoned philosophical framework.

Why did it not? If I were permitted only a one-word answer, I would say: Newton. It was he who demonstrated the full potential of iron rule–governed science with a sweep and force that his contemporaries could not ignore.

In that case, to what happy circumstances do we owe Newton? Was it simply good fortune that Newton came along when he did? Is it possible that in 1642, after millennia of untiring but fruitless play, Mother Nature's great genetic lottery hit the jackpot, assembling Newton's peerless genome like a row of bananas rewarding a last-gasp gamble on a Las Vegas slot machine—that by chance, at an exceptionally fortuitous historical moment, just the right sequence of nucleic acids was strung together to manufacture a mind capable of exploiting to its utmost the magic of the iron rule? Or perhaps more than genes are needed to make a Newton; perhaps it takes the right mix of personal, educational, and social experiences. Either way, a stupendously lucky cosmic accident occurred in mid-seventeenth-century rural Lincolnshire to make the Scientific Revolution not only thinkable, but winnable.

That is intellectually thin gruel; luck is neither a nutritious nor a delicious explainer. There has got to be more to say. Something else was present in the seventeenth century besides Bacon's incipient methodological prescriptions and Newton's idiosyncratic intellect, something that made the irrationality of the iron rule—its rejection of all philosophical, theological, and aesthetic considerations in scientific argument—seem more tolerable, prudent, even civilized than it had ever appeared before. The key to explaining the advent of science, then, is to track down this neutralizer of the rule's intrinsically repellent nature.

EUROPE, once Luther's "Ninety-Five Theses" went out, was forever to be spiritually divided—the new Protestant versus the old Catholic faith; citizens, rulers, and territories of opposing religions living uneasily side by side. Violence followed soon enough: popular revolt in Germany in the 1520s; English dissidents burned at the stake in the 1550s; the desecration of churches in the Low Countries in the 1560s; full-scale war in France from then through the end of the century.

It was to get worse. In 1618, the Protestant nobles of Bohemia, rightly fearing for their religious freedom, spurned their Catholic king Ferdinand—famously by ejecting his representatives from an upper-floor window of Hradčany Castle in Prague—and appointed a Calvinist prince in his place. The deposed Ferdinand soon after assumed the station of Holy Roman Emperor, thereby becoming the ruler of a motley but powerful confederacy of Central European states, some Catholic and some Protestant. He marched on Bohemia to take his revenge.

These were the opening moves of the Thirty Years' War, the conflict that launched Descartes's short, uneventful military career and devastated Europe.

Although the war began over a matter of religious toleration, it soon grew into far more than that: a war between the Bourbon dynasty that ruled France and the Habsburg dynasty that ruled Spain and the Holy Roman Empire; a war by which the German princes asserted their rights against the Holy Roman Emperor; a war of territorial aggrandizement by mercenary captains and Scandinavian kings. Europe was decisively transformed. From a complex, shifting web of loosely affiliated fiefs connected by religious observance and dynastic alliance, it became a mosaic of nation-states held together by something quite new: patriotism and a national interest. As C. V. Wedgwood wrote in her classic history of the war:

> The terms Protestant and Catholic gradually [lost] their vigor, the terms German, Frenchman, Swede, [assumed] a gathering menace. The struggle between the Habsburg dynasty and its opponents ceased to be the conflict of two religions and became the struggle of nations for a balance of power.

Within these nation-states, religion was less and less a prime organizing principle:

> It was not that faith had grown [weaker] among the masses; even among the educated and the speculative it still maintained a rigid hold, but it had grown more personal, had become essentially a matter between the individual and his Creator.

For civic purposes, what mattered now was that you were English or French. That you were Anglican or Catholic was your private concern. When the Peace of Westphalia brought the war to an end in 1648, this

Figure 11.3. Life during the Thirty Years' War: *The Hanging*, from *The Great Miseries of War*, by Jacques Callot, 1633.

separation of political and religious identity was far from complete. But the new organizing principles were clear to, and eagerly debated by, the intelligentsia.

A citizen of a late seventeenth-century European nation-state had to live under two distinct regimes: the spiritual regime, within which they were subject to the will of God, and the civic regime, within which they were subject to the decrees of the monarch or the parliament. If a person cannot serve two masters, then these two roles—law-abiding citizen of the state and obedient servant of God—must not clash in what they require either in action or in thought. The law of the state ought not to prescribe religion, and God's law ought not to override the prerogatives of the state: coinage, taxation, public security, conscription. Or as Isaac Newton wrote, "The laws of God & the laws of man are to be kept distinct." The price of peace in Europe was a permanent bisection of the moral domain into nonoverlapping spheres of obligation, holy church and sovereign nation, each with its proprietary principles and its separate networks of duty and just desert.

To Aristotle, the iron rule's proscription of philosophical and theological argument would have seemed arbitrary and indefensible. To a seventeenth-century mind, by contrast, the rule was asking no more of its adherents than the contemporary political and religious settlement: that a dedicated intellectual space be reserved for its exercise, inside which only a strictly constrained set of principles would be allowed to govern the course of reasoning. Partition—civil and spiritual—was the order of the day. The iron rule's cognitive and logical partition might well have seemed distinctly, fashionably, glamorously, irresistibly "modern." Once the central characteristics of iron rule–governed science were put on the table by radicals such as Bacon, they stood to be taken more seriously than at any other time in human history.

Even after the glorious discoveries carved out by Newton's narrowly empirical method were published in the *Principia* in 1687, it was far from inevitable that his strategies would be universally adopted by other natural philosophers. In that case, the Newtonian opus might have stood alone, towering over the subsequent centuries as Aristotle's philosophy did for so long, admired but unparalleled. That is not what happened: thinkers took up Newton's project, continuing to deepen its fundamentals and to extend its range—developing theories of heat, light, electricity, and the structure of matter—within the same austere framework used by the master himself.

For Newton to be emulated so effectively, his successors had both to discern something like the iron rule at work in his research and to adopt it unreservedly in their own. The discerning was perhaps not too much trouble; Newton lent a hand by adding his famous methodological remarks—"I do not feign hypotheses"—to the second edition of the *Principia* in 1713.

For a thinker to commit to the iron rule in their own research, however—that required finding a way past the blatant irrationality of the rule, its exclusion of philosophical and theological reasoning regardless of merit. Even someone sincere in their intention to follow the rule might easily slip back into the forbidden modes of thought in those cases where they seemed relevant and compelling—as, for example, Boyle did advocating atomism in the years before Newton.

The morally compartmentalizing tenor of the times must have helped Newton's followers stick to their resolve. But even as Newton's work became an object of universal admiration, something else was emerging that was surely even more helpful: the recognition of a division between public argument in the official channels of scientific communication, properly policed by the iron rule, and private thought, in which philosophy, theology, and beauty are allowed free rein. The division is not there in Newton himself; as best we can tell, he applied the dictum that "only empirical evidence counts" to his inner thoughts about mathematical physics as much as to his published pronouncements— such were the workings of his theatrical psyche, separating and isolating methods like sparring characters in the great drama of inquiry. For those equipped with more ordinary minds, however, it would be far easier to commit to the iron rule if its application were required only in argument and not in thought. After Newton, then, the uptake of the iron rule was aided enormously by its confining the scope of its edicts against subjectivity and nonempirical argument to scientific journals and the like, leaving private reasoning unconstrained—a self-imposed limit to the power of the rule that is an essential part of modern science as we know it.

An explanation of science's advent must therefore account for the iron rule's restricting its attention to public argument alone. The full story is long and complex and is indeed still unfolding as standards for scientific objectivity and methods of "sterilization" evolve. That the Royal Society's members conceived of their house journal—founded in 1665, just as Newton began to think about the nature of gravity and light—as a repository of "naked fact" surely played a part. But that in itself does not account for the carving out of a private space in the scientist's mind where the rules dare not intrude. It is possible to glimpse in a celebrated seventeenth-century crisis of conscience, however, the social conditions that favored this final step in the construction of the knowledge machine.

In 1675, Isaac Newton was facing social and professional disaster. Seven years earlier he had been elected a Fellow of Trinity College, and Fellows were required, within seven years of their election, to take holy orders—to be ordained as priests of the Anglican Church. That fate Newton could not contemplate with good conscience. His close study of scripture had convinced him that Jesus Christ was not equal in status to, but was rather created by, God the Father, a denial of the doctrine of the Holy Trinity, the three-in-one nature of the Christian God from which his own college took its name. He was, from the Church of England's point of view, an Arian, a heretic. From his own point of view, it was the church that was in error—an error that he traced back to fraudulent emendations of the scriptures in the fourth and fifth centuries. He could not sully his soul by becoming an agent of a perverted religion; he could not become an Anglican priest. But in that case he would have to relinquish his fellowship, something to which he seemed at one point in 1675 sadly resigned.

Not only would he lose his income; if the reason for his stepping down became known—if his colleagues at the university or in the Royal Society guessed, as well they might, that heresy lay behind his reluctance to take holy orders—then he would be ostracized, "branded a moral leper."

Newton was saved by King Charles II. In response to a plea by someone important at Cambridge—perhaps Isaac Barrow, Newton's predecessor in the Lucasian professorship—a royal dispensation was granted: no holder of the professorship would have to undergo ordination. Provided he was willing to remain silent about his beliefs, then, Newton could continue in his position at Trinity as long as he wished. And so he did, leading two intellectual lives in parallel: one as an orthodox Fellow of Trinity College, publicly conforming to the precepts of the Church of England; one as an anti-trinitarian heretic, privately subscribing to beliefs that would outrage respectable society.

Newton's dilemma illustrates a conflict that arose repeatedly through the seventeenth century. Because the partition of civic and religious life brought by the wars was imperfect and incomplete, and because political and religious borders did not always coincide, it was common for the ambitious or successful to find themselves obliged to say or perform in public things that they privately disdained.

Secrecy and circumspection have, of course, always been useful for surviving despotism or fanaticism. Testaments to the importance of keeping your mouth shut abound:

> Do not befriend kings until you have trained yourself to obey them in matters that are reprehensible to you; to agree with them on matters that you disagree with; to appraise things according to their desires rather than to yours.

So advised the Egyptian civil servant Shibāb Al-Dīn Al-Nuwayrī in early fourteenth-century Islamic Cairo. What made seventeenth-century Europe different from other places at other times was that the divide between the public and private was to a great extent normalized, to a great extent socially accepted. Public speech codes were made very clear and strictly enforced, yet not seriously expected to constrain private thought. The alternative—perpetual religious conflict—was too awful to countenance.

This was the practical beginning of the modern liberal ideal of religious tolerance. It was not the ideal we have today: neither Trinity College nor English society tolerated Arian heresy. But the king's relaxation of Newton's ordination requirement reflects an understanding that a certain degree of play is needed to ensure that the intellectually and spiritually diverse institutional components of the early modern state mesh smoothly, along with a willingness to accept outward conformance to the rules as a sufficient qualification for good citizenship. Thanks to its nascent liberalism, then, the European seventeenth century stands apart from other places in history in its explicitly discussing, regimenting, and occasionally celebrating the segregation of strictly policed outward expression and unfettered personal opinion. This goes some way toward explaining how those who built on Newton's science found it natural and easy to conform to the iron rule's strictures in public scientific debate, even as their minds trafficked in ideas they dared not speak aloud.

THE THINKERS WHO invented science were already experts, theoretically and often practically, in the subdivision of thought, both into autonomous domains—the political and the spiritual—and into the public and private spheres of expression. Such distinctions were hardly unknown in other parts of the world and in other times, but in seventeenth-century Europe they acquired an urgency and, far more important, a legitimacy among the privileged classes that was of historically unprecedented magnitude. The seventeenth century was ready for the iron rule, and when the rule appeared, it made the most of it—it made modern science.

IV

SCIENCE NOW

CHAPTER 12

Building the Scientific Mind

ೞ

How ordinary humans are transformed into modern scientists
through a morally and intellectually violent process

H OLES—EVERY DAY for six weeks, digging holes. That was the
agenda for advanced undergraduate students in soil science at the
University of California, Berkeley, on a field trip described by Hope
Jahren in her memoir *Lab Girl*:

> The average person cannot imagine himself staring at dirt for lon-
> ger than the 20 seconds needed to pick up whatever object he just
> dropped, but this class was not for the average person.

After the daily holes were dug, the scientific work began:

> Every feature of every hole was subject to a complex taxonomy,
> and students would become proficient in recording each tiny crack
> made by each plant root using the official rubric developed by the
> Natural Resources Conservation Service.

Such was the training of these aspiring scientists. Their careers might not be any more thrilling: they were for the most part headed not to universities but to "practical land-management jobs" where their expertise in soil analysis would be applied in much the same way as it was during that hot dusty summer in California's Central Valley.

Why did they put up with it? Scientists embrace the tedium of empirical work, I have written, because the iron rule says that's what it is to do science. Every generation of future scientists, however, has to be convinced to subject itself to the rule. For many, perhaps, it is enough that the rule is part of their job description. The undergraduate soil scientists might have seen their holes simply as a route to a steady salary. For that, they would do whatever the *Keys to Soil Taxonomy* told them to do.

But not everyone will be so easy to persuade. The grave-robbing neuroscientist Santiago Ramón y Cajal and the theoretical physicist Steven Weinberg both record a period of youthful excitement about philosophy. In order to win their Nobel Prizes, they needed to turn away from such temptations and to submit to a regime, as Cajal put it, of "indefatigable persistence and enthusiasm for the observation of facts."

It is the job of science educators—high school teachers, professors, mentors, lab directors—to help them do so, taking a sensibility like Cajal's or Weinberg's, thirsty for knowledge from every discipline in every domain, and training it for a lifetime of argument in which appeals to philosophy, to religion, to beauty, to anything but empirical testing, play no part.

The irrationality of the iron rule throws up a formidable obstacle to the process. Consider that reformed philosopher Weinberg again. He is a notable advocate, in his popular writings, of the probative power of beauty: "We would not accept any theory as final unless it were beautiful." This declaration reveals an apparent conviction that ugliness is a decisive falsifier, a beast so rare that even Popper failed to make a confirmed sighting. Yet at the same time, Weinberg accepts without comment the scientific orthodoxy that aesthetic appeals should find no place in the official channels of scientific persuasion—that in public scientific argument, only empirical testing counts.

No doubt like many of his colleagues, he thinks that it is sufficient, for the needs of science, that beauty exert its influence behind the scenes, whispering in the ear of any scientist who grasps its intimate connection to truth. But why, of all things, should beauty be condemned to creep around in the shadows? How did such a thoughtful writer make his peace with this perverse demand?

The answer, I believe, is that modern scientists' fidelity to the iron rule is inculcated by a method other than persuasion.

To INVESTIGATE THE dissemination of the iron rule, you might take a scientific approach, carefully monitoring the classroom activities and office hours of science teachers in the schools and universities, tagging along on their hole-digging field trips, and eavesdropping on the advice that senior scientists impart to their underlings over the bustle of the lab bench. I propose, by contrast, a more philosophical approach. Ask yourself: if you had to get the iron rule into the heads and hearts of the next scientific generation, how might you proceed?

Imagine, for example, that you are back in Atlantis. Your mission to bring the intellectual and material benefits of modern science to the Atlanteans is well underway. You have convinced them to abandon their rhyming conception of explanation and to begin to care about causes, valuing hypotheses not for their words' sweet sound but for their ability to tell a detailed story about the production of observable phenomena.

Next you must bring them to care *only* about causal explanation; to agree to conduct their scientific conversation, their collaborations, disputes, and mediations, exclusively in terms of explanatory power; to publicly renounce many of what they sincerely regard as their most powerful arguments, their philosophical, aesthetic, and religious reasons to believe. You must, in short, induce them to make the irrationality of the iron rule their own.

Perhaps you could explain to them that nonempirical thinking—appeals to metaphysics, God, beauty, and the like—is bad for science. But some of them, like Steven Weinberg, won't believe you, and indeed there is good empirical evidence for the efficacy of nonempirical reasoning in science. A sense of theoretical symmetry or elegance, in particular, has been crucial at various points in the history of scientific inquiry.

Better, then, is an iron rule that doesn't apologize, doesn't explain—that simply puts its foot down and insists on compliance. Science educators might, in the words of one of Thomas Kuhn's more controversial papers, recognize a vital role for dogmatism in imparting scientific habits of thought.

To implant a dogma in a critical young mind is not, however, entirely straightforward. Even if the iron rule goes, by professorial edict, unquestioned in the classroom, there will remain the temptation to cheat after hours, philosophically or aesthetically circumventing its directives. To curb such desires, you might steal two ideas from the rough trade of politics—morality and simplemindedness. The moral strategy: convey a sense that thinking that draws on philosophy or faith or beauty violates the sanctity of science. The simplemindedness strategy: deprive your students of the ability to think philosophically, theologically, or aesthetically at all.

Moral training first. You want your novitiates to feel bad, feel guilty, feel corrupt pursuing anything but empirical reasoning in science. You cannot create a moral intoxicant from nothing and expect it to have psychological force. Draw, then, on ingredients already known for their power to bring the ego to heel: purity, humility, restraint, asceticism. School your pupils in the following precepts:

The purity of scientific reasoning must not be adulterated by nonempirical strains of thought.

The scientist approaches nature with the utmost modesty; they do not presume to dictate terms, but rather respectfully listen to what she has to say.

Philosophical speculation in science is self-indulgence, abandoning the discipline of empirical testing for the extravagance of speculation.

The scientific life is one of sacrifice; the scientist is prepared to give up almost anything of value to gain knowledge of the natural world.

Don't try to rationalize these rules; let them stand alone as moral absolutes; they are to be preached rather than to be justified or explained.

To that end, equip Atlantean science with a cadre of spiritual leaders to extol the value of empirical testing and to dismiss or denigrate other routes to knowledge of the natural world, such as philosophical reasoning. These moralizers should hold up science as an institution with a privileged connection to the truth: not just the best method of inquiry, but the only proper method of inquiry.

So indoctrinated, the Atlantean scientists will perhaps be better able to resist their higher urges—to forgo the grand project of drawing together all the threads of human reason, philosophical, artistic, empirical, political, and spiritual, to form a greater whole. It is easiest to follow the dictates of a moral system, however, in the absence of temptation.

Thus, the complementary element of Atlantean science education: to foster simple minds. Why furnish a future scientist with all the protocols of philosophy, the paraphernalia of art, only to tell them not, under any circumstances, to be seen putting these accoutrements to use? They make the student a more fully realized person but quite possibly a worse scientist. Better not to take the risk. Therefore, remove such things altogether from the scientific syllabus. To build the surest, purest empirical minds, equip students of science with only empirical modes of reasoning and only empirical knowledge; give them the capacity for empirical thought alone. The belief that they follow the one true way—rightful and uncorrupted—should be satisfaction enough.

THE DYSTOPIAN INDUCTION of young Atlantean scientists into some high church of empiricism couldn't bear any resemblance to science education in the real world today—could it?

It is in fact not so difficult to discern a moral quality in contemporary science's empiricist strictures. Most striking to those in my own profession is the scientific high and mighty's continual denunciation of philosophy.

Stephen Hawking and Leonard Mlodinow's 2010 book *The Grand Design* opens by announcing:

> Philosophy is dead. Philosophy has not kept up with modern developments in science, particularly physics. Scientists have become the bearers of the torch of discovery in our quest for knowledge.

Apparently, Hawking himself had not kept up with modern developments in philosophy: there are philosophers who specialize in the prospects of string theory, the implications of cosmology, and the vicissitudes of quantum gravity, some working at Hawking's own university at the time he wrote these words. But the facts were irrelevant. Hawking was delivering a sermon, not a seminar. The take-home message: philosophical thinking has no place in science.

The astrophysicist Neil deGrasse Tyson mused along similar lines in a 2014 podcast interview:

> My concern here is that the philosophers believe they are actually asking deep questions about nature. [But in fact they are not] productive contributor[s] to our understanding of the natural world. . . . So, I'm disappointed because there is a lot of brainpower there, that might have otherwise contributed mightily, but today simply does not. It's not that there can't be other philosophical subjects, there is religious philosophy, and ethical philosophy, and political philosophy, plenty of stuff for the philosophers to do, but the frontier of the physical sciences does not appear to be among them.

Philosophy, in other words, should stick to gods, morals, and government.

The physicist Lawrence Krauss clarified in an interview for *The Atlantic*:

> Philosophy is a field that, unfortunately, reminds me of that old Woody Allen joke, "those that can't do, teach, and those that can't teach, teach gym." And the worst part of philosophy is the philosophy of science. . . . It has no impact on physics whatsoever.

Numerous outraged commentators have fought back on philosophy's behalf: "It's shocking that such brilliant scientists could be quite so ignorant." But to contest the fairness or veracity of scientists' antiphilosophical remarks is to miss the point. Their function is exhortatory: *Young scientists, shun philosophy and all its ways.* Hawking, Tyson, and Krauss are not cultural commentators with any knowledge of or interest in philosophy; they are holy men chanting empiricist invocations, laying down the credo that shapes and inspires their order of truth seekers.

As a last example, take this 2014 tweet from the biologist Richard Dawkins:

Philosophers' historic failure to anticipate Darwin is a severe indictment of philosophy. Happy Darwin Day!

On the surface, Dawkins's remark makes no sense at all. No one anticipated Darwin; that is precisely what makes for a scientific breakthrough. For every great discovery, there is a parade of past thinkers who failed to make it—a parade that includes philosophers, but also everyone else. In the wake of what appeared to be a staggeringly ill-considered outburst, many criticized Dawkins for not thinking straight. But this was not supposed to be thinking; it was supposed to be incantation. Science soars only once it jettisons the dead weight of philosophy; congregants, lift up your empirical hearts and rejoice!

BESIDES MORALITY, the other underhanded Atlantean strategy for cultivating the iron rule was miseducation. That method, too, seems not to have been neglected in the real world.

All through the 1990s, in universities across the West, there blazed a high-end cultural and intellectual dispute known to its protagonists as the "science wars." A familiar array of questions constituted the battlefield: whether scientists' professional decisions were affected by personal allegiances or cultural background; whether there is any objective element in scientific reasoning; whether science reveals facts about an observer-independent reality.

In the most famous engagement of the conflict, the physicist Alan Sokal submitted to the postmodernist journal *Social Text* an article investigating the liberating potential of the quantum physics of gravity. After the piece was published, Sokal revealed that it was a hoax, an experiment intended to answer the question: "Would a leading North American journal of cultural studies . . . publish an article liberally salted with nonsense if (a) it sounded good and (b) it flattered the editors' ideological preconceptions?" And he reported the result: "The answer, unfortunately, is yes." The topic of the issue in which Sokal's essay appeared was, aptly, the science wars themselves. Afterward, the "Sokal affair" was widely discussed in newspapers and magazines and spawned a number of books by Sokal and others.

Strangely enough, scientists themselves did not seem aware that a battle was raging around them. Stephen Jay Gould recounted his efforts to clue them in:

> Tell most scientists about the "science wars"—and I have tried this experiment at least fifty times—and they will stare back at you with utter disbelief. They have never encountered such a thing, never read anything about it, and don't care to interrupt their work to find out.

What could explain their ignorance? Do scientists consider their humanist colleagues' concerns to be unworthy of their attention?

No, writes Gould; the explanation is not arrogance but "philistinism lite." Most scientists do not notice anything much outside science. Or more exactly, while they may pay attention to the news, to sports, to music, to church, to their families, they know and care little about the many forms of thought that intersect with scientific inquiry—such as philosophy, theology, and the history and sociology of science. It is not that they understand the nature of the claims made, that they appreciate the relevance of these other intellectual pursuits but, like compartmentalizers such as Newton and Whewell, they plug their ears for the greater good of science. Rather, they are barely aware that these ways of thinking exist.

It is straightforward to isolate empirical thought from what's in the other compartments when the other compartments are empty. And so the Baconian prescriptions of empirical science—refer in public discourse to the data, all the data, and only the data—are supremely easy for the average scientist to follow. They don't know how to do otherwise. As E. O. Wilson remarks, "So many scientists are narrow, foolish people."

As he does not remark, that is the secret to their success. An inability to think outside the box funnels all of a scientist's mental and physical and emotional energy into the box itself and thus into the empirical investigation of a single question, the exploration of a single structure, the fabrication of a single substance. It is through this concentration that the iron rule gives the knowledge machine its laser-like power to cut to reality's quick.

You might justly say of many great contemporary scientists what Tolstoy's Prince Andrei says in *War and Peace* about military brilliance:

> A good commander not only does not need genius or any special qualities, but, on the contrary, he needs the absence of the best and highest human qualities—love, poetry, tenderness, a searching philosophical doubt.

It seems apt that Andrew Schally, who together with Roger Guillemin won the Nobel Prize for discovering the structure of the hormone TRH, compared his scientific endeavors to Napoleonic warfare.

THE PROGRAM OF MORALIZING and miseducation that I have sketched in this chapter is not prescribed by the iron rule itself. The rule imposes narrowness on scientific argument and dialogue as it passes through official channels, but as you now know well, it puts no constraints whatsoever on a scientist's private thoughts and feelings.

Indeed, among successful scientists there are extraordinary people who conform to the iron rule in their technical writing but are in no way narrow. They have wandered far outside the boundaries of conventional scientific training and delight in what they find there. They pay little attention to Hawking's or Dawkins's provocations; they take gleeful advantage of the latitude allowed by the iron rule and follow their own tastes and inclinations wherever they lead.

Among them you might encounter "philosopher-scientists," such as Albert Einstein and the eighteenth-century physicist, mathematician, and social thinker Émilie du Châtelet (who translated Newton's *Principia* into French), or thinkers as familiar with history and literature as they are with the technical apparatus of their craft, such as Stephen Jay Gould and Murray Gell-Mann. They may write books about beauty in nature that celebrate the ideas of Pythagoras and Plato, like the theoretical physicist Frank Wilczek. They may champion the aesthetic and moral importance of natural diversity, like Rachel Carson and E. O. Wilson. They may explore the implications of human cognition for life and history, like the psychologists Alison Gopnik and Steven Pinker. These thinkers are, precisely because of their expansive interests, far more likely to be known to most readers than the great, silent, scientific majority upon whose minds scientific training has fixed ponderous iron clamps.

The clamps are, however, the norm. They are the twentieth and twenty-first centuries' standard mechanism for turning out new scientists, instilling the iron rule by psychological stratagems rather than by enlightenment or persuasion.

What a contrast this makes with the first modern scientists, the inheritors of the Newtonian method in the seventeenth and eighteenth centuries. They followed the iron rule, but they were not its captives. Engaging in public argument, they would perform the role of "the empiricist" in the same way that Newton performed the roles of the mathematical physicist, the alchemist, and the scriptural exegete. Outside that context, in their private scientific thought worlds, they would

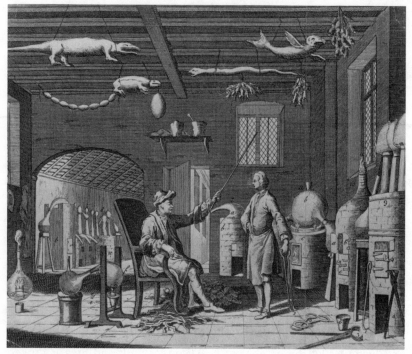

Figure 12.1. The empiricist.

put away the rule and open their minds to whatever seemed pertinent and compelling.

The European seventeenth century excelled in producing minds ready to pull off this theatrical feat. Deeply experienced with exacting or arbitrary rules of public engagement, such minds were able to play their scientific parts to perfection, becoming—while strutting the empirical stage—for all intents and purposes deaf to a chorus of urgent philosophical demands and numb to their most deeply held spiritual beliefs. They thrived within the harshest official strictures, inhabiting their characters not reluctantly or reservedly or half-heartedly or merely dutifully, not rebelliously and not (too) subversively, but with a fervent desire to succeed, taking the framework seriously without forgetting its mere conventionality, putting all of their heart into a role without letting the part engulf their soul.

For that sort of temperament, the ongoing, everyday public performance of narrow and unrelenting empiricism does not pinch off the interior philosophical, spiritual, and aesthetic conduits. Although such a performance must be the centerpiece of the scientific life, what is excluded from the performance does not atrophy but waits patiently in reserve, ready to take back control behind the scenes and between one show and the next.

Oppression and bloodshed were the conditions under which these protean, multifarious minds evolved. We are—most of us—fortunate not to live in such dangerous and trying circumstances. Across the richer half of the globe, humanity enjoys a great degree of tolerance and openness in matters of religion, politics, and philosophy. Consistency between outer actions or words and inner beliefs can be attained without sanction, even without great effort.

And such consistency should be among our highest goals, we moderns tend to believe. Authenticity is a cardinal virtue of our age:

Resolve to abide by your own deepest promptings. (D. H. Lawrence)

To be nobody-but-yourself—in a world which is doing its best, night and day, to make you everybody else—means to fight the hardest battle which any human being can fight. (E. E. Cummings)

In this ever-changing society, the most powerful and enduring brands are built from the heart. . . . The companies that are lasting are those that are authentic. (Starbucks founder Howard Schultz)

To conform to such precepts makes for a purer and more perfect realization of our ideal of what it means to be human. But at the same time it produces minds that are ill suited to the theatricality and normative compartmentalization that keep the iron rule in its proper place. The highest expression of liberal democracy undermines, in other words, the cognitive, emotional, and social skills needed to maintain a science that is both widely receptive—tuned in to the universe at every frequency—and intensely empirically focused.

The focus is essential; without it, the knowledge machine loses its traction on the world. So we build scientific minds that are empirical in thought as well as in words. That makes for scientists better able to live authentic lives, realizing their empirical values personally and professionally, in private and in the wider world. It empowers science to make all lives better materially and intellectually. It makes peace with what we consider to be a far better polity: we would much prefer to live in a liberal democratic regime than to negotiate the complexities and rigidities of seventeenth-century civic, religious, and social life. From an intellectual and cultural standpoint, however—and perhaps a moral standpoint, too—it makes for a less agreeable science. The knowledge machine, in its contemporary realization, is highly effective at advancing human goods, but it is not a high expression of what is humanly good.

CHAPTER 13

Science and Humanism

ⓧ

The fullness of humanistic thought against the poverty of
scientific thought; the effectiveness of scientific thought against
the impotence of humanistic thought

"LET US BE DRIVEN, O Fathers, by those Socratic frenzies which lift us to such ecstasy that our intellects and our very selves are united to God." So wrote the Renaissance humanist Giovanni Pico della Mirandola (1463–1494) in his *Oration on the Dignity of Man*, sometimes called the manifesto of the Renaissance.

By what means was humankind to achieve a knowledge so exalted that its intellects, its selves, would be "united to God"? By all means possible. Pico himself read Latin and Greek and studied Hebrew, Arabic, and Aramaic. He drew on Plato and Aristotle, on Islamic philosophers such as Averroes, and on the Talmud and kabbalistic texts, in an attempt to find, in the integration of his philosophical, religious, and mystical sources, the quintessence of knowledge. The *Oration* was intended as a preface to a great meeting of minds that Pico hoped to stage in Rome, in which 900 theses, drawn from ancient Greek, Christian, Jewish, and Moslem sources, would be debated. In the course of the dialogue, the wisdom each contained would be extracted, distilled, and then blended to create an exhilarating concoction that would transport the human mind to the greatest heights of understanding.

We can put aside Pico's conception of the end of inquiry as a mystical union with God while finding in his project—as have his many modern readers and admirers—a portrayal and a celebration of a humanist ideal of knowing. This ideal upholds an integrating conception of knowledge, according to which the surest path to the most important truths brings together all sources of insight: philosophical, spiritual, poetic, mathematical, experimental, as well as everyday experience of the world. It is a route to enlightenment mapped out by many other Renaissance thinkers—such as the early sixteenth-century Swiss physician Paracelsus mentioned in Chapter 10, who combined a commitment to empirical experiment in medicine and chemistry with a devotion to the allegorical thought of alchemy, along with the idea of a grand symmetry governing the universe at both the astronomical and human scales.

A hundred years later, we encounter the same all-embracing ideal in Descartes. He took for his subject matter just about everything: the causes of motion, the structure of the universe, the emotions, the nature of thought, God, mathematics, the foundations of knowledge—philosophy, psychology, physics, theology, and more. Dashing back and forth, his reasoning interleaved and tightly wove these separate topics. His physical theory hinged on his philosophical argument that empty space is impossible. How could he be so confident? His philosophy of knowledge assured him that careful reasoning based on clear ideas could not go wrong, in part because God is responsible for planting those ideas in our heads. Why think that there is a God, and for that matter, why think that God wants us to be enlightened rather than merely ignorant and awed? Descartes gave two philosophical arguments for the existence of a benevolent God. His physics, then, is built on his philosophy of matter, which depends on his philosophy of knowledge, which depends on his theology, which depends in turn on more philosophy.

In his religiously conventional French rationalism, Descartes was quite different from Renaissance figures like Paracelsus and Pico. Rigorous and systematic metaphysics rather than syncretic magical hermeticism was the form of his deepest thinking about the nature of the material world. But his answer to the question "By what means shall we know the world?" was the same as Pico's: *By all available means.* From metaphysics and mathematics, from introspection and observation, a single, coherent theory of the world should be drawn. That is the humanistic way.

The term *humanism* has been used to mean many things. Secular humanism, a modern idea, signifies a renunciation of all gods, all religious sources of meaning. That is not what I have in mind; for both Pico and Descartes, God and the spiritual plane are objects and foundations for knowledge. In another sense, humanism is a historical phenomenon confined to the Renaissance and concerned with the resurrection of classical learning. Although humanism in my sense is amply represented in Renaissance thought, it is far wider in scope. Aristotle, for example, is a paragon of my sort of humanism, mingling philosophical argumentation with observation, explanatory speculation, and a little theology.

If Aristotle, Pico, and Descartes are all fundamentally humanistic, what great thinker is not? The personification of science himself: Isaac Newton.

Newton was in one sense a supreme Renaissance man, his interests quite as broad as Descartes's. He was not only an empirical scientist but a mathematician, an alchemist, an interpreter of scripture, and, like Descartes, a metaphysician using philosophical argument to understand the nature of space and matter. Unlike Descartes, he quite deliberately failed to integrate these investigations. Each went forward under the power of its proprietary techniques, without assistance from the others. In his compartmentalization of inquiry, he practiced—indeed, he pioneered—an approach that is a stark negation of humanism's synthesizing ethos.

Were Descartes a university, it would be a rambunctious and vibrant place, spilling into the hallways and the stairwells. Every member of the faculty would read, discuss, and argue about the work of every other. The physicists and the philosophers, the theologians and the psychologists, would participate equally in a shared discourse on the principles that rule the world.

Were Newton a university, you would not hear a sound; the common room would be thick with dust. Each faculty member would be found at all times shut up in their office or lab, pursuing their own researches by their own means, reading just those books that speak directly to their own subject matter, writing only books answering to the same specification. They would meet once a year to discuss parking and the budget for coffee.

And which would be the greater institution? The principles of rationality and the humanistic spirit give the same answer: the Cartesian university, the very incarnation of the most human qualities in its vibrant sociality and open-mindedness. Experience says otherwise. It is the Newtonian university's taciturn specialization that is the better route to knowledge. Whatever is lost through detachment and disregard for the grand view of life is more than recompensed by the narrow, tightly focused beam that searches out the diminutive but telling fact.

The bitter fate of humanist thought has been to see its glorification of the full, unifying intellectual potential of the human mind eclipsed by the immensely greater contribution to our knowledge of the natural world made by the lean scientific spirit—to comprehend the meagerness of what is found beyond the golden gate of the imagination when measured against the riches brought back through the low and shameful gate of experience.

THE NEWTONIAN UNIVERSITY is an allegory of Newton's mind, not a genuine institute of research and higher learning. It has something in common, nevertheless, with the enterprise of contemporary science.

This is due in part to the iron rule's governing the way that science talks to the world. Leaf through an assortment of scientific journals and you will find a tidy array of self-contained compartments, aloof and insular, an embodiment of the narrow empiricist code.

At the same time, the iron rule—I'll say it again—leaves scientists quite free, in their private lives and interior deliberations, to range over whatever theological or philosophical or aesthetic territory they wish to explore. Even if a certain Newtonian quietude prevails in the great public spaces of science, the back rooms and connecting passages are wide open to untrammeled Cartesian hubbub. Indeed, scientific discovery relies to a not inconsiderable extent on this furtive openness, which has allowed thinkers such as Murray Gell-Mann, D'Arcy Thompson, and Albert Einstein to use their aesthetic and philosophical senses in the search for extraordinary theories.

These great scientists were exceptional in more than one way. Not only were they brilliant and imaginative; they also succeeded in evading the deadening effects of contemporary science's preferred method for imposing the iron rule, a method that starves scientific novices of nonempirical knowledge and undercuts nonempirical habits of mind.

The standard product of this system is an empiricist all the way down, an individual who not only in their public writings but also in their private thinking takes a "scientific attitude" that is directly opposed to the humanistic attitude. The scientific attitude demands tangible evidence. It scoffs at philosophy and is uneasy with a sense of beauty or meaning that cannot be put into words. It finds fulfillment in direct, unemotional, indeed colorless expression of ideas and arguments. It transmutes the iron law of explanation into a leaden law of scientific thought.

ASIDE FROM THE OCCASIONAL outbreak of plague, life as a university student in medieval times seems to have been much like it is today: a steady rotation of classrooms, taverns, and last-minute study sessions. The books, however, were rather different. There was, in the twelfth century, no *Norton Anthology of English Literature*. But there was Martianus Capella's *On the Marriage of Philology and Mercury*, "the standard schoolbook of the Middle Ages."

Written as the ancient Roman Empire's power in the Mediterranean crumbled—most likely sometime between the Visigoth Alaric's sack of Rome in 410 CE and the invasion of Martianus's native city of Carthage by the Vandals in 429—*On the Marriage* describes a celestial union conducted in the palace of the gods Jupiter and Juno, floating beyond the outermost planets. The bridegroom Mercury stands for eloquence and the art of persuasion, embodied in the study of grammar, logic, and rhetoric, while the bride Philology stands for the love of learning and inquiry into the workings of the world, embodied in the study of arithmetic, geometry, music, and astronomy. The wedding is therefore a synthesis of three "humanities" and four "sciences," making up the seven liberal arts—who serve as bridesmaids of a sort and who occupy the greater part of the book with learned presentations of their domains of knowledge, jointly portraying an ideal of the educated mind as fluent in modes of thought that span the sciences and the humanities.

Numerous writers have lamented what appears to be, in the era of contemporary science, a de facto divorce. The English chemist and novelist C. P. Snow famously declared in 1959 that science and the humanities had diverged to the point that they formed two distinct cultures, each largely unconscious of the other's subject matter and methods. Snow deplored the situation; his ideal was a thinker conversant with, if not expert in, all branches of human knowledge and invention. The same sentiment is expressed in Stephen Jay Gould's book on science and the humanities (*The Hedgehog, the Fox, and the Magister's Pox*); and Gould might have been surprised to come upon, had he lived to read it, a similar outlook in the culminating sentence of his sparring partner E. O. Wilson's *Meaning of Human Existence*:

> If the heuristic and analytic power of science can be joined with the introspective creativity of the humanities, human existence will rise to an infinitely more productive and interesting meaning.

Who could fail to be moved by this vision? It is, I regret to inform you, too lovely to be true.

There are two missteps in Snow's famous lecture. The first is to talk of the modern science we have built, our contemporary knowledge machine, as a culture. It is not even a subculture. It is a social practice, and one that is carefully cultivated in the gardener's sense, but it bears little resemblance to what an anthropologist would call a culture. It has norms—a moral code, if you like—but the central tenet of that code enjoins scientists, in their professional lives, to shun all entanglement with broader intellectual or spiritual matters. It is socially embedded, and like any social unit it has its traditions and its quirks. Otherwise, it is more like a rule book, a corporate headquarters, or a military unit than a form of life. Its function is not to enable a certain way of being in the world, but to suppress the impact of our being in the world on our knowing of the world.

Snow's second misstep—and Wilson's too—lies in suggesting that science would flourish if scientists knew and cared more about the rest of existence. Quite the contrary: their obliviousness is the greatest guarantee that they will follow without deviation the empirical path laid out by the iron rule. All the power to improve the world that Snow and Wilson hoped for lies along that path.

According to a story in Genesis, the creation of humanity was an act of inspiration:

> And the Lord God formed man of the dust of the ground, and breathed into his nostrils the breath of life; and man became a living soul. (Genesis 2:7)

The creation of the knowledge machine was just the reverse. Its engineers brought it into being by sucking the air out of the chambers of the mind—the philosophical air, the theological air, the air of beauty, the humanistic spirit. It was this act of deprivation that manufactured the void in which empirical inquiry can most effectively divine the facts.

In Joseph Wright of Derby's most famous painting, *An Experiment on a Bird in the Air Pump*, a group of onlookers—both rapt and

Figure 13.1. William Blake, *Elohim Creating Adam* (detail), 1795.

Figure 13.2. Joseph Wright of Derby, *An Experiment on a Bird in the Air Pump*, 1768.

appalled—watch a bird die slowly as the oxygen is extracted from the bubble of glass in which the bird is imprisoned. Scientific instruments are spread across the table like tools of torture. The room's only source of light is obscured by a goblet in which floats some nameless horror. Through the window faintly shines a wan, gothic moon. The experimenter, with his long and gray Newtonian locks, shows no mercy.

The bird in the painting is doomed. But science is a different kind of creature. It loves the bubble; it thrives in the vacuum. What would kill it is to let in the scents, the commotion, the delights of the outside world.

CHAPTER 14

Care and Maintenance
of the Knowledge Machine

ℰℐↃ

Research has given us hope that all
Shall be well and all manner of thing shall be well
Till the moment it's not. It's not.
"POEM," MAUREEN MCLANE, *SAME LIFE*

NATURAL PHILOSOPHY BEGAN, as far as we know, when Thales of Miletus conjectured, in the sixth century BCE, that the elemental stuff of which the world was made was water. At that time, water was the lifeblood, if not of the universe, then certainly of the local economy: Miletus, controlling a harbor at the center of a vigorous trading network in the eastern Mediterranean, was in Thales's day said to be the wealthiest city in the world. Modern Miletus is a ruin, an empty place many miles from the sea. Centuries of deforestation and overgrazing allowed colossal quantities of silt to be washed into the Meander River and down into the bay that gave Miletus its access to the ocean. Year by year, the bay became shallower and then began to disappear. It is now the dry, dusty plain over which the remaining stones of Miletus preside.

The last of the sea withdrew from Miletus around 1500. Five hundred years later, it came to New York City. Hurricane Sandy brought a 14-foot storm surge that flooded low-lying areas throughout the five boroughs. In the lobbies of the Financial District, waves lapped around the

Figure 14.1. The remnants of the city of Miletus in the nineteenth century.

security cameras and the turnstiles, the filing cabinets and the flatscreen monitors. The power went out for a week.

The inundation, we are increasingly aware, was a visitation from the future. By the year 2050, according to one recent report, rising sea levels may have forced 300 million coast dwellers to abandon their homes.

The reasons for Miletus's demise could be counted out in terms of herds of goats and passels of timber; the underlying cause, however, was the Milesians' and their neighbors' unending desire for more. To satisfy their appetites, the wooded slopes of the interior were cleared and transformed into farmland; a naturally dry climate and relentless erosion did the rest. The richer Miletus grew, the faster it brought on its own destruction.

We moderns have our appetites, too. The consequences might seem quite as fearful. But even if the knowledge machine, by supercharging industry, is in part to blame for the rapidity with which our proclivities and our proliferation have degraded our habitat, it offers us, at the same time, our best chance of salvation. Science will, if anything can, show us how to satisfy our wants without draining the earth of all that it needs to support life; it may also, if we treat it right, show us how to repair some of the damage that's already been done.

So how do we nurture the complex of individuals, institutions, and instrumentation that constitutes modern science so as to get from it the knowledge that will allow us to continue to live the good life? We must do two things. The first is to set the agenda, using grant money and government initiatives to steer a sufficient number of scientists toward questions that matter. The second is to secure the smooth operation of the knowledge machine itself, ensuring that it purrs—efficient, responsive, dynamic, and strong—as we career through the twenty-first century and beyond.

To address these matters in full would be a vast undertaking, another volume. With some regret, I put the question of agenda setting to one side. To make a start on the question of the maintenance of the knowledge machine's internal engineering will, however, be a satisfying way to conclude this book, taking a final, backward glance at my attempt to resolve the Great Method Debate. Let me begin by asking what my inspirations and adversaries, Karl Popper and Thomas Kuhn, might advise.

THAT ARCH-METHODIST, Karl Popper, predicated science and indeed civilization on a simple truism: a theory that makes mistaken predictions must itself be wrong. This, the principle of falsification, was the torch that Popper held high to light the path out of the darkness that had haunted his youth, away from social breakdown, mob violence, and mechanized slaughter.

A logical rule cannot be improved upon. But the scruples of the people who employ it most certainly can. To that end, Popper believed that it was imperative to fan the critical spirit in science. Inside the white coats, behind the heavy spectacles, there must be installed a pitiless sensibility that would overlook no predictive defect—that would stand ready to call out falsehood, without regard to how hallowed, how plausible, or how beloved the faltering theory might be. A distinctly inhuman elevation of logic over loyalty, according to Popper, is what we should implant and cultivate in the human minds that make up the knowledge machine.

We've seen an insuperable technical obstacle to the Popperian plan. Logic alone cannot definitively falsify a theory; plausibility rankings—that is, subjective estimates of the likelihood of various auxiliary assumptions—will always play a role. Indeed, it is in virtue of these rankings that scientific opinion changes, ultimately flowing toward the truth in the course of Baconian convergence. A purely logical machine, then, will be a poor scientist.

Might we not strive, nevertheless, for scientific minds that are as unbiased, as unprejudiced, as unprovincial as possible? Certainly, a science of fierce, critical, selfless minds would be an extraordinary thing to behold—like the government of the just or the communion of the holy. But we'll see it only in our dreams. The knowledge machine is made of human beings, not an angelic host. We need a science that tolerates—or even better, harnesses—human frailty.

In any case, a science dominated by critical spirits might revoke the irrational elements of the iron rule, summoning philosophical and aesthetic reasoning back into scientific debate. The resulting enterprise would look less like modern science and more like the natural philosophy of old, like Descartes altercating across the centuries with Aristotle—and would perhaps be no more effective.

More of the critical spirit in our politics—yes. In our bureaucracies—yes. And in society at large—absolutely, yes. The world of human affairs could use far more than its meager allotment of Popperian rationality.

But to the exotic creature that is contemporary science, it would bring argumentation in suffocating excess.

THOMAS KUHN'S HANDBOOK for a healthy science makes a rather different set of recommendations. The great Kuhnian insight is that what distinguishes modern science from ancient and medieval science— that is, from "natural philosophy"—is not a superior logical tool kit or advanced technology, but a special form of social organization, the "paradigm." This all-encompassing methodological framework provides the moral, intellectual, and emotional support that is necessary, as Kuhn wrote, for a scientist to "investigate some part of nature in a detail and depth that would otherwise be unimaginable."

Kuhn's prescription for effective scientific inquiry is therefore to bolster the framework, to exalt the paradigm. All scientific research programs in a given domain must, at any one time, be corralled within a single, shared set of rules, shaped in accordance with a distinctive explanatory and methodological creed. Scientists working within such programs must be committed so deeply to the paradigm that they simply take its rightness for granted, finding alternative frameworks to be preposterous or even inconceivable.

The Kuhnian scientist is, as a consequence, not a free, critical spirit— and Kuhn thinks that this lack of freedom and criticism is essential to good science. A Kuhnian program of science education, aiming to optimize the power of the knowledge machine, will discourage autonomy, free thinking, and resistance to the status quo.

Kuhn is right about something fundamental to the operation of scientific inquiry—the importance, when going after empirical evidence, of passionate intensity—but his regime, like Popper's, asks for too much from its personnel. As 50 years of the sociology of science have shown, it is simply not realistic to suppose that scientists will follow the paradigm as slavishly as Kuhn envisaged in *The Structure of Scientific Revolutions* or—more important still—that they will maintain complete faith in its validity under any circumstances short of its total collapse. The pungent mix of strength and weakness in the human will, sometimes assiduous and sometimes inattentive, and in the human spirit, alternately devoted and defiant, will see to that. A contingent of ideal Kuhnian scientists, working in strict adherence to protocol, is more redolent of a military parade than of a university department—and indeed, it is far easier to keep legs in lockstep than minds.

So much for the old methodists, Popper and Kuhn. The new methodism proposed in *The Knowledge Machine* suggests three essential ingredients for a thriving science.

The first is fighting spirit. I don't mean Popper's critical spirit, scrutinizing the theoretical landscape from disinterested logical heights. What I have in mind is in far more plentiful supply: partial, self-interested ambition. Such ambition need not be low-minded; its interest in seeking out the truth and advancing human happiness may well be sincere. Nor need it be combative or mean—a great athlete can be full of grace. But it must be ready to play the game to win.

The fighting spirit must then be caged within the iron rule. The nature of the game is thus defined: from its players it will elicit the kind of evidence—arduous and expensive to produce—that hones the knowledge machine's sharp edge, and it will store that evidence securely for thinkers in the centuries ahead. The human race provides fighting spirit in abundance; the iron rule, by contrast, was hard to come by, because its demands are to all appearances contrary to reason. Indeed, as revealed by its war on theoretical beauty, they are in the fullest sense irrational. Nevertheless, the rule's dominion over all forms of inquiry into nature is now well established. "Only empirical testing counts" has come to feel normal, even rather boring.

Perhaps a little too boring. Thus, the third and last of the knowledge machine's needs, which might also be the most difficult, in our day, to satisfy. It is to leave science alone, that is, to resist the urge to tinker, to make science more current, more flexible, or, for that matter, more sensible.

The pressure to "improve" science could come from any direction—from funding bodies, technology companies, or political actors. It could even come from scientists themselves. One striking instance of this last possibility—of a mutiny against the iron rule from within the knowledge machine—may be seen in the arguments that have swirled around the status of string theory for the past 20 years or so.

String theory (more properly "superstring theory") has for decades been proffered as a promising "theory of everything," providing a unified framework to explain both gravitation, currently handled by Einstein's general theory of relativity, and the other fundamental forces, currently handled by the Standard Model of particle physics. String theory has many seductive features, but it is extraordinarily difficult to test in the way mandated by the iron rule. Certain experiments that would prove decisive would require, it is said, detectors the size of the planet Jupiter or particle accelerators as large as our galaxy.

In string theory's defense, some physicists have advanced what look like amendments to the iron rule. One suggested revision would allow string theory's unity, beauty, and coherence to count in its favor, not only in scientists' private thinking—that is already permitted—but in the course of official scientific argument. This amounts to a suspension of the iron rule's decree that all official debate go by way of empirical testing. The scientists who back the proposal are sometimes said to be advocating "post-empirical physics." Should such an endeavor be allowed?

It is an enticing prospect. I have said myself that aesthetic reasoning can yield deep insights into the principles of nature, and so that its exclusion from scientific argument violates a fundamental precept of rational thinking. All string theory's champions are asking, then, is that the scientific method be given a logical upgrade.

And yet—the method has a function that transcends its lamentable logic. It is there to make profound, exacting experiment happen. Allow aesthetics into science, and the pressure to measure will subside. I can't say for sure that it will lead to disaster; the risk, however, is forbidding.

Do not, then, meddle with the iron rule. Do not tamper with the workings of the knowledge machine. Set its agenda, and then step back; let it run its course.

BUT I HAVE NEGLECTED to consult the radical subjectivists—those thinkers who hold that the machine, science, has no special method, no fixed operating procedure.

The sociologists Harry Collins and Trevor Pinch, paragons of radical subjectivism, compare science to a golem—an automaton made from inert matter and actuated by magical words written on paper and pushed into its mouth, supernally powerful but nearly impossible to restrain. "It will follow orders," they write, "do your work, and protect you from the ever threatening enemy. But it is clumsy and dangerous. Without control, a golem may destroy its masters with its flailing vigor."

When I suggest that science will better flourish if allowed to go its own way, then, Collins and Pinch might think that I have given a bad answer to the wrong question. What we should ask is: how can science be reined in?

The archetypical golem was brought into being in the late 1500s by the leading rabbi of Prague, Judah Loew ben Bezalel, to protect the city's Jewish population. According to one account, when it began to run amok, the rabbi had to take back the animating incantation—*emet*, the Hebrew word for truth—from the golem's mouth. His creation turned to dust.

In the legend, you might discern the radical subjectivist agenda: take away science's claim to absolute truth and it will be tamed. A more careful reading of Collins and Pinch's metaphor shows, however, that neutralization of the knowledge machine is very far from their aspirations. They take the golem to be a "bumbling giant," a "creature of our art and our craft," whom for all its faults "we should learn to love for what it is." The aim of their book is not to bring science to heel, but to understand how it works, to see how its knowledge is made.

That is my project, too. I do not, as you know, endorse the understanding of science proposed by subjectivists such as Collins and Pinch. They have overlooked the iron rule; consequently, they have missed the scientific method. Restore the method to its proper place at the heart of the machine, however, and the image of the golem retains its force.

The golem of legend was made of clay and brought to life by a magic word. The science golem's raw material is people, organized and empowered by the iron rule. (It is bronze, not iron, but otherwise the sculptor Eduardo Paolozzi's transmogrification of Blake's Newton, shown in the frontispiece to this book, is an apt portrayal of a man made knowledge machine.)

The golem of legend had a mind of its own. Science is made up of many minds—those of a multitude of scientists, each interpreting the evidence in the light of their own culture and education, their own taste and inclination, and acting accordingly. Yet at the center of this multiplicity sits something further, the public arena in which scientific debates unfold. That we can understand as a theater of consciousness belonging to the golem itself. The golem's memory is the archive of observation and experiment; its trains of thought are the arguments published by disputing scientists. Such reasoning as goes on in the golem's head is therefore incomplete and conflicted. But at the same time, it operates at a remove from the human concerns of the people of which it is made.

Having waited so long and struggled so hard to create this thing, how can the human race best benefit from its existence? Point it in the right direction and let it go. Empirical science is a dull beast, but feeling no pain and knowing no fear, it can do something that we with our refined minds and delicate sensibilities, so easily distracted, cannot. Snout to the ground, it is oblivious to the political and personal concerns of individual scientists, leaving their cultural baggage and petty self-interest behind.

This magnificent obtuseness may show us the way to sustainable happiness. It comes, though, with a cost and a compromise. The cost is irrationality: the creature abandons the baggage without regard for its value. The compromise is ambivalence: until Baconian convergence is attained, science makes no judgments about what the evidence shows, whatever individual scientists may believe.

Science's irrationality we'll tolerate, even welcome, if it lends the golem greater strength. For a society in search of answers, the ambivalence is rather more daunting.

The weather is getting wilder. Populations are on the move. Exotic diseases—Ebola, AIDS, SARS, MERS, Zika, COVID-19, which is rampaging as I write—are vaulting from animals to humans every generation. Technology is decreasing in size and growing in power like an ever more tightly sprung trap. We've pampered and praised the knowledge machine, given it the autonomy it has needed to grow. Now we desperately need its advice.

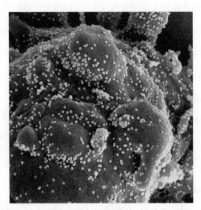

Figure 14.2. A cell heavily infected by SARS-CoV-2, the coronavirus that causes COVID-19

With enough evidence in—with Baconian convergence achieved or at least well on the way—there will exist a consensus among scientists that is functionally equivalent to science's speaking with a single voice. But at the moment when some critical concern assumes maximum urgency, the evidence is often patchy and provisional: different scientists, bringing different plausibility rankings to bear on the same question, will express diverse opinions. The voice of the golem in these cases sounds not like a harmonizing choir but like the babble before the music begins: the clamor of a thousand crosscutting conversations.

To make sense of the cacophony, we must find an interpreter. In the dire case of climate change, the foremost interpreter is the IPCC, the Intergovernmental Panel on Climate Change. Coordinated by the United Nations, the IPCC gathers scientists from around the world in working groups that contribute to assessment reports issued every few years. The aim of the reports is to summarize the state of scientific knowledge concerning the climate; among other things, they assign confidence levels to hypotheses, perhaps attaching "medium confidence" to one and "very high confidence" to another, and they assign likelihoods to particular events—such as a 3-degree increase in average global temperature by 2050 or a 5-inch increase in sea level by 2100—using expressions such as "more likely than not," "likely," "very likely," and so on. (In the same way, the UK, US, and other governments have convened committees of experts during the COVID-19 crisis to extract predictions, as best they can, from a bewildering array of conflicting epidemiological models.)

Such a panel, for all its expertise and hard work, cannot determine what science says. Science holds no determinate views. The IPCC's numbers are created, as all such numbers must be, by infusing the scientific record with a set of plausibility rankings. Although the IPCC aims to use a range of rankings that reflect, in some sense, the center of mass of scientific opinion, they are subjective all the same: they are not derived from the objective evidence, but are rather what must be added to the evidence to induce it to begin to talk. It follows, says Stephen Schneider,

Figure 14.3. Hurricane Sandy floods the Brooklyn-Battery Tunnel in New York City, October 29, 2012.

a lead author of several of the IPCC reports, that "if we care about the future, we have to learn to engage with subjective analyses."

We do indeed. Interpretation requires a worldview. Kuhn taught that science is blind to any worldview other than the prevailing paradigm. He overstated its intelligence and sophistication: it is blind to worldviews altogether. The unstinting focus that results is what makes science so inexorable a stalker of knowledge. To fathom all the knowledge it finds, however, we must bring our subjectivity to the task, looking into the monster's mind with human eyes. In this one crucial respect, the radical subjectivists are right.

THE KNOWLEDGE MACHINE opened in the darkness of prehistory. Civilization's sun rose, bringing literature and law, temple domes and proscenium arches, and the more abstract pleasures of mathematics and philosophy. Science's sun, meanwhile, remained deep below the horizon. To one surveying the cultures of the ancient world, there was no glimmer to suggest that anything like modern science would arise. So it continued for centuries, millennia. Empires came and went; each left its enduring aesthetic and intellectual gifts to humankind, but there was no science.

At a stroke, the Scientific Revolution changed everything. Science's sun seemed to have appeared, not on the horizon, but at its zenith, as the fierce genius of Newton and his lieutenants glistered in the heavens. It burned far hotter than had even the sun of civilization. Our sultry, teeming, denatured planet is its consequence—as are our increasingly long, comfortable, amusing lives.

Galileo yearned to know the nature of light. "I had always felt so unable to understand what light is," he wrote to a friend, "that I would gladly have spent all my life in jail, fed with bread and water, if only I was assured that I would eventually attain that longed-for understanding." Less than four hundred years later, thanks to Isaac Newton, James Clerk Maxwell, and Albert Einstein—along with many others—we have that knowledge. The light of science calls out for the same understanding. In *The Knowledge Machine*, I have given you the truth as I see it.

Science is not light; it is not promulgated by a star. Nor is it a golem, a glass slipper, a neurasthenic bird, or a coral reef. It is not, indeed, a machine. It is a social institution. It could not be brought into existence by a celestial body or by a magical incantation. Inquirers had to give the rule that constitutes the scientific institution to themselves. But the iron rule is a peculiar mix of power and perversity. Logically, it is beyond the pale. It would take an exceedingly long time for social, political, and moral conditions to twine themselves into a perspective from which the rule would seem to be an acceptable idea, fit to enter the halls of inquiry. Now we know. And because of the iron rule, we can go on knowing, more and more. Let us hope that knowledge saves us.

ACKNOWLEDGMENTS

I couldn't possibly list all the writers, readers, teachers, and students who played a role in the genesis of *The Knowledge Machine*, even if I could tease apart the tangled lines of influence and figure out who had prompted, or gently undermined, which ideas. Two of the most important influences, however, have been my former colleague Peter Godfrey-Smith, who helped me to appreciate better the history and sociology of science and with whom I first taught Kuhn, Popper, and many of the other figures in this book, and Philip Kitcher, whose work inspired my first investigations of the social structure of science and who has been a reliable source of insight and support ever since.

Many friends read and commented on various parts of the manuscript: Joy Connolly, Erik Curiel, Rebecca Goldstein, Leanna McLennan, Ashleé Miller, Marco Nathan, Katie Tabb, and J. D. Trout. I thank them, along with Trevor Teitel, for his help in searching the physics literature for nonexistent references to beauty, and Maya Jasonoff, for a bracing excursion through the space of all conceivable titles. Thanks also to the Kavli Foundation and to Steve Hall and Gary Marcus for an inspiring but only too brief education in the art of science writing.

This book would have been quite different, and far inferior, without the exacting attention of two editors: Bob Weil, my editor at Liveright, and Jessica Loudis. Bob's finely judged blend of patience and exasperation, exerted over five long years of writing and revision, was more than any writer could have reasonably expected. I am deeply grateful. It is equally difficult to imagine making it through that long period of gestation—roughly three elephants' worth—without the support and advice (not always concerning writing) of my agent, Chris Calhoun.

As for the business end of the book, I have much appreciated the care and attention of all the people at Liveright—in particular, Gabe Kachuck, Phil Marino, and Marie Pantojan, to name those I worked with directly—my scrupulous copyeditor, Janet Greenblatt, and my indefatigable, unsurpassable image researcher, Trish Marx.

Finally, my thanks to the Magnusons for the use of the Moog, where the book began and prospered.

GLOSSARY OF NOVEL TERMS

BACONIAN CONVERGENCE The process in which the outcomes of empirical tests in the long run bring about, for some given question, consensus on the sole theory able to explain all that has been observed—the true theory.

EXPLANATORY RELATIVISM The thesis that every age or school has its own standards for explanation and understanding. One age's standards, and thus its explanations, will make little or no sense to the thinkers of another age.

IRON RULE OF EXPLANATION The rule demanding that all scientific arguments be settled by empirical testing, along with the elaborations that give the demand its distinctive content: a definition of empirical testing in terms of shallow causal explanation, a definition of official scientific argument as opposed to informal or private reasoning, and the exclusion of all subjective considerations and nonempirical considerations (philosophical, religious, aesthetic) from official scientific argument.

METHODISM The view that science's truth-finding power is explained by a special technique or form of organization, a distinctive way of dealing with empirical evidence, the "scientific method."

PLAUSIBILITY RANKING A scientist's level of confidence that a hypothesis or other assumption is true.

RADICAL SUBJECTIVISM   The view that, because the outcome of scientists' reasoning and argument depends so strongly on individual tastes and goals, institutional preconceptions and politics, and social needs and mores, science has no fixed, rigorous procedure for learning about the world. It follows that there is no objectively superior "scientific" way of thinking, no "scientific method."

SHALLOW CAUSAL EXPLANATION   Explanation in which an event or phenomenon is explained by deriving it from causal laws, other causal generalizations, and background conditions, without any supporting account or philosophical justification of the causal process undergirding the laws.

THEORETICAL COHORT   A theory combined with the sort of further assumptions about the world, including assumptions about apparatus and experimental conditions, that are required in order for theories to have observable implications.

TYCHONIC PRINCIPLE   The fact that the most promising rival theories of the structure of the world make their disagreements manifest in small observable details and only in such details. Baconian convergence on true theories is therefore possible, and only possible, by way of great amounts of complex, highly accurate observation.

NOTES

Some of the sources and resources mentioned here are available to anyone online; some are rather difficult to get hold of if you don't have a university affiliation. I recommend checking authors' web pages for "preprints" (that is, near-final versions of papers whose definitive versions may be behind paywalls). Preprints of my own papers are on my website at http://www.strevens.org. A helpful resource for philosophical topics is the Stanford Encyclopedia of Philosophy (http://plato.stanford.edu).

INTRODUCTION: THE KNOWLEDGE MACHINE

2 **to its stock of knowledge, scarcely at all:** There are honorable exceptions: mathematical astronomers in Babylon and Greece developed theories that accurately predicted the movements of the sun, moon, and planets, and the Greek thinker Archimedes made substantial discoveries in hydrostatics, or the science of floating, immortalized in his wild discoverer's cry "Eureka!"

6 **"There was no such thing as the Scientific Revolution":** The first sentence of Shapin's book *The Scientific Revolution*. The artful discordance between Shapin's title and his first sentence illustrates two ways to use the term *Scientific Revolution*. On the one hand, it is the name for a certain sequence of events—the observation of Jupiter's moons, the discovery of the circulation of blood, the articulation of Newtonian physics—whose occurrence is a matter of historical record. On the other hand, and more contentiously, it is the name for a presumed new intellectual or social movement that precipitated those events. Shapin writes the social history of one while denying the existence of the other.

6 **"still shrouded in darkness":** Feyerabend, *Science in a Free Society*, 73.

CHAPTER 1: UNEARTHING THE SCIENTIFIC METHOD

13 **a condemnation of totalitarianism:** Popper, *Open Society and Its Enemies*, published in 1945.

13 **perhaps the only human activity:** Popper, *Conjectures and Refutations*, 293.

13 **his philosophical masterpiece:** *The Logic of Scientific Discovery* was published in German in 1934; it first appeared in English 25 years later, in 1959.

14 **"incomparably the greatest":** In a BBC radio interview, cited by Magee, *Popper*.

14 "the war years and their aftermath": Popper, *Unended Quest*, 13. The quote that follows is from p. 32.

15 "freedom is more important than equality": Popper, *Unended Quest*, 36.

15 "I remember only": Popper, *Unended Quest*, 37. The quote that follows is from p. 38.

17 "there is no intellectual difference": Russell, *History of Western Philosophy*, III.I.17.

18 "If the redshift": Quoted approvingly by Popper in *Unended Quest*, 38.

19 "Scientific theories, if they are not falsified": Popper, *Unended Quest*, 79.

19 there is no point in trying to gather evidence: Popper did think that scientists should accord a certain respect to theories that have withstood severe testing. He said that such theories were "well corroborated," but held to his anti-inductive view that we have no more reason to believe a well-corroborated theory than a theory that has not been tested at all.

20 "Let our conjectures . . . die in our stead!": Popper, "Natural Selection and the Emergence of Mind," 354.

20 "There is no more to science": Cited by Magee, *Popper*, 9. See also Mulkay and Gilbert's "Putting Philosophy to Work," a sociological study of Popper's influence on scientists' methodological theory and practice.

20 "I learned from Popper": Eccles, "Under the Spell of the Synapse," 162.

21 the anatomist Solly Zuckerman: Nicholas Wade describes the episode in *The Nobel Duel*, p. 54. The brain in question supposedly lacked a channel between the hypothalamus and the pituitary that, according to Harris, was essential to ferrets' coming into heat, which Zuckerman's ferret had nevertheless done. Harris pointed out the many ways in which this one-off case might be misleading. He won the argument and is now regarded by many as the "father of neuroendocrinology" for his theory of hormone-driven communication within the brain.

22 "I understood at last the problem of induction": The story is told by Ronald Giere in his book *Science without Laws*. He attributes it to the philosopher Andreas Kamlah, who, he reports, "knew neither of its origin nor any evidence that it might be true." In my experience, such anecdotes are almost always urban legends, but in the spirit of inductive skepticism, I decline to disbelieve this one on those grounds.

22 Morris recalls that Kuhn: As related in *The Ashtray*. Morris writes, "I had imagined graduate school as a shining city on a hill, but it turned out to be more like an extended visit with a bear in a cave" (p. 9).

22 "a neurotic, insecure young man": The quotes are from the interview with Kuhn in *The Road since Structure*, respectively pp. 280 and 276.

23 "Suddenly the fragments in my head": Kuhn, "What Are Scientific Revolutions?," 9.

24 they would find it incomprehensible: Kuhn talks about the grip of the paradigm in many different ways. Sometimes, especially early on in *Structure of Scientific Revolutions*, the paradigm is merely a set of assumptions that scientists take for granted—albeit assumptions that are strictly socially policed. (Scientists who dare to dissent from the paradigm "are simply read out of the profession, which thereafter ignores their work" (p. 19).) Later in the book the paradigm is said to cause "an immense restriction of the scientist's vision" (p. 64), and by the end of the book scientists with different paradigms "practice their trades in different worlds" (p. 149). In my exposition of Kuhn, I emphasize his stronger statements about the paradigm's hold on scientists' minds.

26 arguably the first scientific revolution: On the question of whether Greek astronomy was sufficiently scientific that the transition from Ptolemy to Copernicus con-

stituted a lower-case scientific revolution, Kuhn was a little vague. The same is true for the transition from special creation to Darwin.

28 **"Normal science . . . is predicated"**: Kuhn, *Structure of Scientific Revolutions*, 5, 52.

28 **some visionary "deeply immersed in crisis"**: All quotes in this paragraph are from Kuhn, *Structure of Scientific Revolutions*, 90.

29 **"conversion experience"**: Kuhn, *Structure of Scientific Revolutions*, 150.

30 **"When paradigms enter"**: Kuhn, *Structure of Scientific Revolutions*, 94.

30 **Perhaps if you had two brains**: Would your two brains understand one another well enough to negotiate a winner? Kuhn suggested that there is no common methodological, conceptual, or linguistic framework that they could use to discuss their differences—a predicament he named "incommensurability." For many years he worked on a definitive treatment of the topic. It never appeared.

30 **the Aristotelian and the Newtonian live in different worlds**: That is the thrust of the controversial Chapter 10 of *Structure of Scientific Revolutions*.

30 **"as in political revolutions"**: Kuhn, *Structure of Scientific Revolutions*, 94.

30 **"mob psychology"**: Lakatos leveled this charge in his long and influential 1970 essay "Falsification and the Methodology of Scientific Research Programmes," 178.

32 **later paradigms tend to have more predictive power**: This is not to say, as Kuhn emphasized, that a new paradigm can predict everything forecast by its predecessor. Predictively speaking, it may be two steps forward and one step back—but that still adds up to one step forward.

34 **the 1977 Nobel Prize**: Guillemin and Schally split the prize with Rosalind Yalow, who was honored for a different but related achievement. An engaging account of the contest between Guillemin and Schally can be found in Nicholas Wade's *The Nobel Duel*.

34 **"immense amount"**: Latour and Woolgar, *Laboratory Life*, 120.

34 **"Nobody before had to process"**: Quoted in Latour and Woolgar, *Laboratory Life*, 118.

35 **biologists Peter and Rosemary Grant**: The Grants' work is memorably described in Jonathan Weiner's *Beak of the Finch*. Work on the new species was reported in 2018, in Lamichhaney et al., "Rapid Hybrid Speciation in Darwin's Finches."

36 **Lewis Terman's decades-long "Genetic Studies of Genius"**: Beginning in 1921, Terman recruited California children between the ages of 8 and 13 who scored especially well on a battery of tests, especially the Stanford-Binet IQ test that he had himself developed. His aim was to follow these "gifted" children through their lives, examining the ways in which their "genius" flowered. But it turned out that they were not much more apt to display great creative power than any other children. The assumption that IQ and genius go hand in hand was mistaken. The project continues to this day, tracing the lives of subjects now in their nineties but with rather different research goals in mind.

37 **a summer she spent in Colorado**: As related in Jahren's *Lab Girl*, 73–75.

37 **"You have to believe"**: Quoted in Aschwanden, "Science Isn't Broken," at http://fivethirtyeight.com/features/science-isnt-broken/.

38 **"Only a person such as myself"**: From the autobiographical note Schally wrote upon being awarded the Nobel Prize (https://www.nobelprize.org/prizes/medicine/1977/schally/ biographical/).

40 **I think that Popper and Kuhn are correct**: Were I keeping score, I would give Kuhn extra credit here: he puts more emphasis than Popper on the motivating power of methodological consensus.

CHAPTER 2: HUMAN FRAILTY

42 **by a young Karl Popper:** As related in the previous chapter, the eclipse experiment and other tests of relativity directly inspired Popper's idea of falsification. Thus, not only was the experiment in the Popperian mold; it molded Popper himself.

45 **"the logic of the situation does not seem entirely clear":** Campbell is quoted on p. 78 of Earman and Glymour, "Relativity and Eclipses," my principal source on criticisms of the eclipse experiment.

A defense of Eddington's neglect of the Brazilian astrographic plates is mounted by the physicist Daniel Kennefick in "Testing Relativity from the 1919 Eclipse." Kennefick clearly explains how a "change of scale" would create a systematic error, but he does not address the curious fact that Eddington and his coauthors make no attempt to convince their readers that such a change had occurred, rather than there having been a simple loss of focus that would not create any systematic bias in the astrographic measurements. As far as we can tell, Eddington simply chose the explanation for the blurriness of the astrographic's photos that best suited his goals. I will take up the question of Eddington's omission again in Chapters 3 and 7.

Matthew Stanley's "An Expedition to Heal the Wounds of War" is also largely sympathetic to Eddington's treatment of the data and provides much fascinating historical background. Stanley gives a broader perspective in his book *Einstein's War.*

45 **"the pursuit of truth":** Quoted by Matthew Stanley, "Expedition to Heal the Wounds of War," 64. Stanley argues that Eddington saw the eclipse expedition as a "religious calling" (p. 59).

45 **"the most influential figure in British astronomy":** Earman and Glymour, "Relativity and Eclipses," 71.

46 **According to Karl Popper:** There are other ways in which the episode of the eclipse resists a simple Popperian interpretation. The Brazilian telescopes yielded two conflicting measurements of the bending of light, which poses an obvious problem for falsification as a scientific method: no theory can predict both outcomes; thus, every theory will be falsified. In the next chapter, we will see what Popper has to say about this.

46 **According to Thomas Kuhn:** Should Eddington's eclipse experiment be regarded as taking place within the Newtonian paradigm or the incipient Einsteinian paradigm? Such questions arise frequently when attempting to read history in Kuhnian terms: science has often failed to unfold in the neat fashion described in Kuhn's *Structure of Scientific Revolutions.* As a consequence, Kuhn continued to refine his conception of the cycle of paradigm-governed science, crisis, and revolution after the publication of his most famous book. Most historians of science, however, while they acknowledge Kuhn's enormous importance, have abandoned the attempt to understand the history of science in terms of a series of successive paradigms for this reason as well as the reasons discussed throughout the present chapter. The role of Kuhn's ideas in contemporary history of science is laid out succinctly in Mario Biagioli's short paper "Productive Illusions."

47 **science's biggest names can be seen discarding:** See, respectively, Fisher, "Has Mendel's Work Been Rediscovered?"; Richardson et al., "There Is No Highly Con-

served Embryonic Stage in the Vertebrates"; Holton, "Subelectrons, Presuppositions, and the Millikan-Ehrenhaft Dispute"; Westfall, "Newton and the Fudge Factor" (the quote is from p. 753). In each case, accusations of impropriety have spurred illuminating debates about the culpability of the scientists and the damage done to science—with some writers arguing for little culpability and less damage—but there is scant doubt that a certain amount of deliberate misrepresentation took place.

48 **Eddington's original presentation:** Dyson, Eddington, and Davidson, "Determination of the Deflection of Light."

52 **Pasteur and Pouchet had sparred:** This story is told in Collins and Pinch, *The Golem*, which also contains a brief, accessible, and rather unsympathetic account of Eddington's maneuvers.

52 **a combative and unfair disputant:** A balanced biography that takes the notebooks into account is Patrice Debré's *Louis Pasteur*.

53 **the industry-supported group is considerably more likely:** On soda: Bes-Rastrollo et al., "Financial Conflicts of Interest and Reporting Bias Regarding the Association between Sugar-Sweetened Beverages and Weight Gain." On secondhand smoke: Barnes and Bero, "Why Review Articles on the Health Effects of Passive Smoking Reach Different Conclusions." On drug trials: Cho and Bero, "Quality of Drug Studies Published in Symposium Proceedings." See also Bekelman, Li, and Gross, "Scope and Impact of Financial Conflicts of Interest in Biomedical Research."

54 **a young German scientist and explorer:** Mott Greene's *Alfred Wegener* offers a comprehensive treatment of Wegener's life, which I draw on here.

57 **we might still be waiting:** The evidence that pointed to seafloor spreading as the mechanism by which continents are moved was uncovered by scientists who were not thinking about drift at all—more reason to think that the last thing science ever should do is to slow down or stop.

58 **scientists in the United States:** Naomi Oreskes, in *Rejection of Continental Drift*, surveys the speculations about the mechanism for drift that appeared in and after Wegener's work. Oreskes's explanation of the controversy is somewhat different from mine, although it exemplifies, in its own way, my point about the importance of differing methodological tastes: she explains the US geologists' caution as grounded in a suspicion of grand, theory-driven geology in favor of a bottom-up, inductive, theoretically pluralistic (or even agnostic) approach. I myself interpret the "inductive" stance of the American geologists as comprising, among other things, a demand that each working part of a big, systematic theory be independently evidenced and thus that the drift mechanism postulated by Wegener's theory be not merely possible in principle but be supported by observations that go some way toward demonstrating its existence.

59 **"[science] is beautifully self-correcting":** In a 2016 commencement address at Cal Tech; a transcript is posted at http://www.newyorker.com/news/news-desk/the-mistrust-of-science.

60 **Latour's "knowledge of science was non-existent":** All quotes in this paragraph are from Latour and Woolgar, *Laboratory Life*, 273.

60 **"inscription devices":** Latour and Woolgar, *Laboratory Life*, 51.

62 **"local tacit negotiations":** Latour and Woolgar, *Laboratory Life*, 152.

62 "We were unable to identify": Latour and Woolgar, *Laboratory Life*, 189.

63 "The work they were doing": Zimring, *What Science Is and How It Really Works*, 279.

63 "The natural world has a small or non-existent role": Collins: in "Stages in the Empirical Programme of Relativism," 3. Aronowitz: in *Science as Power*, 204. "the strongest team": in Patricia Fara, *Science: A Four Thousand Year History*, 257. Brassy subjectivist slogans such as these are collected by Alan Sokal at https://scientiasalon .wordpress.com/2014/03/26/what-is-science-and-why-should-we-care-part-i/.

64 hopeless, each and every one: I have been principally interested, in this chapter, in aspects of scientific practice that undermine the methodist approaches of Popper and Kuhn equally. Thus, I have not made much of signs that many scientists actively seek to overthrow the prevailing orthodoxy (see, for example, "We Want to Break Physics" at http://www.bbc.com/news/science-environment-31162725). Such rebel yells are bad for Kuhn but good for Popper. Likewise, I have not dwelt on cases where scientists unthinkingly echo the orthodoxy—bad for Popper but good for Kuhn.

64 science's theoretical as well as its practical success: The question of the relation between science and technology was much discussed in science studies in the 1980s. Opinion eventually settled on what is called the "interactive view," described by, for example, Alexander Keller: "So—has science created technology? The answer must be twofold. One answer is: No, it was not the prime originator. . . . The other is: Yes, modern technology would be impossible without scientific training and comprehension of the nature of things" (in "Has Science Created Technology?," 182). Keller's piece includes an illuminating overview of postwar research on connections between basic science and technological innovation.

64 "Science remains": Shapin, *The Scientific Revolution*, 165. Even Harry Collins, whose radical dictum was quoted earlier, is happy to say that scientists are "the foremost experts in the ways of the natural world" (writing with Trevor Pinch in *The Golem*, 142).

CHAPTER 3: THE ESSENTIAL SUBJECTIVITY OF SCIENCE

66 "some kind of lighter fluid": Quoted in David Grann, "Trial by Fire," *New Yorker*, September 7, 2009.

67 those who contributed to the case against him: Apparently, the people involved in Willingham's conviction still on the whole believe that he was guilty.

73 do the experiment a second time: Popper's requirement is stronger than this: for falsification he requires "corroboration" of a "falsifying hypothesis," which implies the repeated production or observation of the falsifying fact.

73 "would not occur again for many years": Dyson, Eddington, and Davidson, "Determination of the Deflection of Light," 293.

75 "a creating and directing power": The quotations are taken from Smith and Wise's biography of Kelvin, *Energy and Empire*, 645.

75 But the science of Darwinism: Charles Darwin himself suffered a gradual loss of faith. Reflecting about it later in life, he wrote, "Disbelief crept over me at a very slow rate . . . so slow that I felt no distress, and have never since doubted even for a single second that my conclusion was correct" (Darwin, *Autobiography*, 87).

76 Kelvin presumed the original temperature: Also important to Kelvin's estimate, and also measured by Forbes, was the temperature gradient, that is, the degree to

which the earth gets hotter as you dig deeper. Forbes was of course only able to estimate this gradient very close to the surface.

76 **"much nearer 20 than 40"**: Quoted in Dalrymple, *Age of the Earth*, 43.

76 **Darwin stood refuted:** Perhaps even more than Darwinism, Kelvin was concerned to refute the "uniformitarian" geology of the mid-nineteenth century, with its appeal to the "inconceivably vast" lengths of time needed to deposit the strata and to build the geological formations to which geologists' eyes were for the first time being opened. By finding beginnings and endings, it seems, Kelvin sought to restore a narrative arc to the unfolding of time, making a place for both creation and redemption. But he did not neglect the impact of his calculations for the hypothesis of evolution by natural selection: his limit on the earth's age, he wrote with evident satisfaction, "seem[ed] sufficient to disprove" Darwin's hypothesis (cited in Burchfield, *Lord Kelvin and the Age of the Earth*, 85).

77 **"What you get out":** From Huxley's presidential address to the Geological Society of London in 1869, in which he launched his initial attack on Kelvin's dating. Cited in Burchfield, *Lord Kelvin and the Age of the Earth*, 84.

77 **"Natural Philosophy . . . already points":** Quoted by David Lindley, *Degrees Kelvin*, 176.

77 **he was relying on assumptions:** Kelvin's assumptions are explored in Dalrymple, *Age of the Earth*, 46–7. An engaging presentation of the importance of convection in particular may be found in England, Molnar, and Richter, "Kelvin, Perry and the Age of the Earth."

78 **Huxley was right:** The estimates of the age of the sun were also based on false assumptions. Kelvin and the other estimators, Hermann von Helmholtz and Simon Newcomb, had supposed that the sun's energy was due to gravitational contraction; in fact, it is almost all supplied by the then unknown process of nuclear fusion.

79 **only as strong as its weakest link:** A complex chain of reasoning may have a structure that is more sophisticated than a simple sequence of links. At some points, for example, the chain might briefly separate into two strands, each bearing some of the weight, as when a critical assumption is supported by two independent considerations. In these cases, it is not always true that the chain is no stronger than its weakest link. But the more general moral—that reliably estimating the strength of a chain is impossible without reliably estimating the strength of its individual links—does hold true.

81 **the solidity assumption:** Kelvin in fact devoted considerable time and energy to arguing for the assumption of solidity: he considered various ways in which a sphere of molten rock—as he supposed the earth once was—might cool, and he concluded that according to any of these stories, the earth would have solidified early in its history. He did his best, then, to extrapolate from contemporary physical knowledge, but generalizing to an entire planet turned out to be, as Huxley and company had surmised, a step too far.

82 **an unbreakable circle:** The same circles can turn up even in the absence of big theoretical auxiliary assumptions. One complaint about Eddington's interpretation of the eclipse observations was that, in calculating the light-bending angle implied by the Príncipe telescope's measurements, he at one point assumed the correctness of Einstein's theory—the same theory that the measurements were supposed to

test. Some sociologists call the circularity that precludes an independent objective test of auxiliary assumptions "the experimenter's regress." Philosophers call it (I'm sorry to say) "confirmation holism."

83 **The heart of scientific logic is a human heart:** To more precisely formulate these ideas, philosophers have constructed probabilistic logics to capture scientists' plausibility rankings and their role in evaluating evidence. Such logical systems differ in various ways, but they all implement to some extent a simple idea: if your chain breaks, suspect the weakest link.

The most influential probabilistic system was first developed by the eighteenth-century clergyman and mathematician Thomas Bayes (1701–1761). Bayes's approach cannot be wholly correct—it assumes that scientists know all possible theories in advance—but it provides an excellent model of many aspects of scientific inquiry.

In the Bayesian framework, scientists' plausibility rankings for the things that matter—hypotheses and auxiliary assumptions—are represented by numbers between 0 and 1, varying from person to person. From some simple mathematical facts and an equally simple supposition about the way that scientists take new evidence into account, it is easy to show that the significance of a piece of evidence can vary dramatically, depending on a scientist's plausibility rankings. Indeed, one and the same piece of evidence may appear either to support or to undermine a theory, given different rankings. The demonstration abstracts away from the flesh-and-blood reasons why scientists differ in their rankings: it treats aesthetic judgments, politically expedient rationalizations, and raw self-deception in exactly the same way. But in so doing it exposes something that all plausibility rankings, whatever their origins, share: the power to foment disagreement as to what the evidence says, a power that consequently frustrates any attempt to formulate an objective rule for weighing evidence. Some further implications of Bayesian logic are discussed in a note to Chapter 5 ("plausibility rankings begin to converge").

CHAPTER 4: THE IRON RULE OF EXPLANATION

92 **Perhaps, he suggests:** An alternative explanation posits an invisible substance filling even apparently empty spaces, through which heat could flow in the same way that it flows through more everyday materials. That is the transitional "wave theory" of heat, which supplanted caloric theory in the 1830s but then itself yielded to kinetic theory in the 1840s. Apart from the need to explain heat's ability to travel through empty space, the wave theory was motivated by the widespread belief (eventually shown to be false) that heat and light are at bottom the same sort of thing.

92 **The sun's heat:** Besides infrared radiation, a good amount of the sun's heat comes by way of visible light in virtue of a process slightly more complicated than simple absorption. This complication does not affect the following discussion, which is premised on Montague's assumption that light rays and heat rays are distinct kinds of radiation.

92 **These heat rays are not:** As physicists were to discover in the twentieth century, electromagnetic radiation is composed of streams of photons, particles that behave partly in a wavelike, partly in a raylike, manner.

95 **If heat rays are like light rays:** Ordinary glass does in fact block some "heat rays" (that is, some frequencies in the infrared spectrum). A black plastic garbage bag, though opaque to visible light, is transparent to infrared radiation.

96 **find an experiment or observation:** Sometimes testing turns on an observation or experiment that is not quite so revealing: one hypothesis (or rather, cohort) has an opinion about the outcome, but the other does not. The second hypothesis is, in that case, indifferent to what happens; in effect, the observation tests the first hypothesis while the other stands by, benefiting in a passive way if the first fails the test, like a politician taking advantage of an opponent's scandal.

97 **the nub of the scientific method:** As I have formulated it, the iron rule turns on hypotheses' ability to explain. Scientists often say, however, that it is by predictions that theories are tested. Wasn't it predictive differences that distinguished the Newtonian and Einsteinian theories of gravity or, indeed, the caloric and kinetic theories of heat?

Yes it was, but such predictions are typically explanations as well. Einstein's theory not only predicts that light passing the sun will be bent to a certain degree; it also explains why it is bent to precisely that degree. Likewise, Montague's theory of radiation not only predicts that heat will pass through glass without slowing; it explains why that is so. Insofar as this duality of prediction and explanation holds, it hardly makes a difference whether the iron rule is formulated in terms of one or the other.

There are numerous explanations, however, that are not predictions: evolutionary theory, the social sciences, much of geology and economics, and even cosmology explain far more than they predict. Evolutionary biologists can give a rather good explanatory account, for example, of various distinctive physiological features of the human body, but they could never have predicted its form if, impossibly, they were conducting their research 30 million years ago, just as the ape and monkey lineages diverged.

What makes prediction harder than explanation in a case like this is that prediction, but not explanation, requires everything relevant to be known in advance. In explanation, by contrast, you have the many benefits of hindsight. To predict the behavior of a friend or a lover in a sticky situation—that can be a challenge. But in the aftermath, the explanation for what they did is often abundantly clear. Life is like that.

So is the science of life. The biologists of 30 million years ago could not have predicted that modern humans would walk on two legs, because they could not have known for sure either that full bipedalism was developmentally possible in the monkey lineage—that the body plan of humanity's ancestors would allow such a physiological configuration—or that it would be selectively advantageous. From the perspective of now, these uncertainties are eliminated: the possibility and advantages of bipedalism are manifest. Facts discovered after the event go into the explanatory theory like any others, allowing a derivation that would be beyond our reach if what were being explained had not already come to pass.

If the criterion for empirical testing is to regulate the pursuit of historical sciences such as evolutionary theory, then it is explanatory rather than predictive power on which the criterion for empirical testing must be built.

98 **empirical inquiry has seen nothing like this:** The one exception I can think of is

the formation of psychoanalytic schools after Freud—Jung, Adler, and the rest. But this was possible precisely because the psychoanalytic tradition came to reject the procedural consensus, allowing matters other than observable evidence into the debate. In this respect, it is closer to ancient natural philosophy, which begat the Stoic and Epicurean schools, than it is to modern psychology: it is an attempt to frame a worldview, not a mere science.

99 **the hypothalami of 160,000 pigs:** From Schally's Nobel autobiographical note (https://www.nobelprize.org/prizes/ medicine/1977/schally/biographical/).

99 **"He is the Establishment":** Quoted in Latour and Woolgar, *Laboratory Life*, 119. Schally is likely referring to S. M. McCann's unsuccessful work on LRF, according to Nicholas Wade in *The Nobel Duel*. Wade summarizes McCann's own diagnosis of his failure: "His prime strategic error, as he sees it, lay in not allocating more of his resources to the LRF chase. Had he devoted more of his budget to buying hypothalami, had he diverted less of [his] stock of LRF to interesting physiological experiments, perhaps the outcome of the LRF race [would] have been different" (p. 222).

101 **"I *hate* the Standard Model":** For other particle physicists with the same attitude, see "We Want to Break Physics," cited in a note to Chapter 2 ("hopeless, each and every one").

103 **my disagreements with Kuhn:** Let me register one further difference between us. For Kuhn, the "rules of science" are more like a way of life: they tell you not only which moves are allowed, but also what counts as a good move and what constitutes "sportsmanlike behavior" (among other things). The iron rule is much closer to the bare regulations of a sport, that is, to what is laid out explicitly in the rulebook. A move can be legal but inadvisable, a sure losing gambit. Likewise, an empirical test can be technically relevant to the argument, yet inconclusive or redundant—as when the same experiment is repeated over and over with the same result.

CHAPTER 5: BACONIAN CONVERGENCE

105 **Bacon admired the ancient Greek natural philosophers:** Bacon considered early natural philosophers such as Thales and Anaximander to be superior to Aristotle and Plato. To account for the greater fame of the latter, he ventured: "Time (like a river) has brought down to us the lighter, more inflated works and sunk the solid and weightier" (*New Organon*, §1.71).

105 **"round in circles for ever":** This and "a new beginning" from Bacon, *New Organon*, §1.31.

107 **"as innocent as any":** According to John Campbell, *Lives of the Lord Chancellors*, 404.

108 **"Every contradictory instance":** Bacon, *New Organon*, §2.18.

108 **"The quiddity of heat":** Bacon, *New Organon*, §2.20.

109 **Bacon's method provides no objective guidance:** Bacon himself encouraged scientists to develop working hypotheses as they go along, "because truth emerges more quickly from error than from confusion" (§2.20). But he did not propose a procedure for choosing between such hypotheses in the absence of fresh evidence. His own "first harvest" theory of heat turned out to be on the right track, so the issue did not arise in his great worked example.

110 **plausibility rankings begin to converge:** The Bayesian probabilistic logic, which, as I remarked in a note to Chapter 3 ("The heart of scientific logic is a human heart"), shows that short-term divergence of opinion is inescapable, also shows that under certain circumstances, long-term convergence of opinion is inexorable. The mathematics of the Bayesian system proves, in other words, that as evidence continues to come in, scientists starting out with divergent plausibility rankings will in their opinions come closer and closer to each other and to the truth. As the "under certain circumstances" hedge implies, however, this encouraging conclusion must be qualified in certain ways, corresponding to complications remarked upon in the next note.

112 **agreement emerges again and again:** Baconian convergence is not easy, and philosophers have articulated powerful reasons to worry that with respect to some matters, it might not occur at all.

First, in certain domains there might be more than one theory that explains all the observable facts, no matter how many are collected. There would be no Baconian way to choose among these champion explainers.

Second, according to the Baconian ideal, the sheer mass of evidence overwhelms, ultimately, all scientific partisanship and prejudice. But if the evidence is interpreted from a partisan and prejudiced point of view, why be so confident that the evidence will win and partiality will lose? Worse, what if there are false presuppositions about the world hardwired into the human brain? Every scientific thought would be inflected by these presuppositions; you might fear, then, that even the most potent dose of empirical evidence could not possibly expunge them from the gray matter—that they will live on in our theories like conceptual parasites as long as the human race persists.

These worries can be quelled, and the resulting discussion is frequently fascinating, probing some of the most profound issues in all of philosophy. To take on such weighty topics would, however, be too much of a burden at this point in the book. Perhaps I should write a second volume? For now, in lieu of that volume, I have offered a simple empirical argument that Baconian convergence is possible: modern science, though young, has achieved it already many times over.

112 **Written by Sir Francis Bacon:** Ignatius Donnelly, a late nineteenth-century populist US politician who presciently (as it will soon emerge) maintained that the island of Atlantis once really existed, published *The Great Cryptogram* in 1888, arguing that the works of Shakespeare were in fact written by Francis Bacon. The same thesis was advocated by a number of other nineteenth-century literary adventurers. It has fallen, for good reasons, into disrepute.

112 **In a basement in Ohio:** I simplify some of the details of the Michelson-Morley experiment in the main text. The apparatus was slowly rotated, so that all directions of travel relative to the earth's were tested. The beams traveled back and forth a number of times before their arrival time was compared. And the beams were compared by viewing the interference pattern created when they were superimposed.

116 **"forces scientists to investigate":** Kuhn, *Structure of Scientific Revolutions*, 25.

117 **the first great scientist in history:** Armand Marie Leroi, in *The Lagoon*, writes beautifully about Aristotle's copious observations of and theories about nature, emphasizing their scientific spirit: "He invented the science [of biology]. You could argue that [he] invented science itself" (p. 7).

117 **David Lindberg provides a partial list:** In *The Beginnings of Western Science*, 362–4.

CHAPTER 6: EXPLANATORY ORE

123   "no timeless, ahistorical criteria": Dear, *Intelligibility of Nature*, 14.

123   **All things, animal, vegetable:** Aristotle notes that there is a difference between inert matter and animals: inert matter is not genuinely self-propelled, or else a dropped coffee cup would not crash witlessly to the floor but would find some way to perpetuate its fall, perhaps ducking out of a window to get closer to its natural nesting place.

123   **"fell into place":** The quotes are excerpts from a longer passage reproduced in Chapter 1.

126   **his well-tended beard:** Jonathan Barnes writes of Aristotle, "He was a bit of a dandy, wearing rings on his fingers and cutting his hair fashionably short" (in his *Aristotle*, p. 1).

126   **"Physics alone cannot explain life":** Aristotle makes his argument against the materialists in *Physics*, II.8.

127   **It is rather something:** Plants, too, have *psuche*s, according to Aristotle, accounting for their functional, adapted forms.

127   **the form of the animal in its body:** I have taken this formulation from Leroi, *The Lagoon*, 158. For my purposes, it is an adequate and appealing paraphrase of Aristotle's doctrine that the *psuche* is the "first actuality" of the body (*De Anima*, II.1).

129   **the Thirty Years' War would have killed:** On the complicated business of estimating the war's casualties, I have relied on Peter Wilson's *The Thirty Years War*, Chapter 22.

130   **he dreamed three times:** The content of the dreams—heavily condensed and highly interpreted here—was recorded by Descartes and related by his early biographer Adrien Baillet; descriptions can be found in any modern biography of Descartes, from Stephen Gaukroger's rich and highly readable *Descartes: An Intellectual Biography* to Richard Watson's charming, discursive, and somewhat eccentric *Cogito, Ergo Sum*.

132   **a kind of centrifugal force:** As with centrifugal force in Newtonian physics, Descartes understands this tendency not as a genuine force, but rather as an expression of rectilinear inertia—the tendency of objects to continue moving in a straight line unless something holds them back or gets in their way. Objects moving in a circle, such as anything perched on a rock that orbits the sun, will experience this tendency as a palpable tug pulling them outward and away from their circular path. The planets themselves are subject to the same inertia; it is the invisible matter in which they're embedded that prevents them from making a beeline out of the solar system.

132   **"nothingness cannot possess any extension":** Descartes, *Principles of Philosophy*, §II.18.

132   **"It is no less impossible":** From a letter to Mersenne cited by Garber, *Descartes' Metaphysical Physics*, 131.

133   **all other bodily functions:** There is one great exception: conscious thought. Our mental life, believed Descartes, plays out in an immaterial soul that communicates with the body through a small structure in the brain called the pineal gland. However this happens, it cannot be through physical contact—a rather vexing problem

for a philosopher who has just proved beyond any shadow of personal doubt that all material change is caused by collisions.

134 **Aristotle would dismiss Descartes's as hopeless:** Cartesian philosophy would seem to be a prime target for Aristotle's ancient argument against a wholly materialist biology—the argument that the biological world's finely tuned harmony cannot be explained by something as dumb and unruly as impact. Descartes might have attempted to evade this charge by invoking what was for him the religiously orthodox view that God was responsible for bringing biological forms into being through an act of special creation. Aristotle, had he allowed such an event, would still not have been satisfied. Even if the living world were miraculously assembled at a single stroke in all its perfect complexity, Aristotle would contend, that perfection could not last. Like a city abandoned in the jungle, biological structures would slowly deteriorate as small aberrations crept into the mechanisms by which organisms sustain and reproduce themselves. Could God continue to oversee his handiwork, keeping mechanical processes in line and on target to do what's best for an organism? Cartesian mechanics, in that case, would be mere metaphysical theater; the real causation would be God's pulling the strings behind the scenes. Descartes was too pure of a thinker to find such a picture appealing. He insisted that the simple laws of physics were enough to preserve the biological attunement that Aristotle believed could be explained only by *psuche*.

135 **"Truth is the offspring of silence":** Newton's words are quoted in Gale Christianson, *Isaac Newton*, 21. Readers looking for a short life of Newton have two superb choices: Christianson's and James Gleick's.

136 **"barbaric physics":** Leibniz's vituperative words are in an unpublished essay titled "Anti-Barbarus Physicus," tentatively dated 1710–1716 (in *Philosophical Essays*, 312).

137 **how Kuhn spins the history of gravity:** Kuhn, *Structure of Scientific Revolutions*, 103–5.

137 **a postscript:** Newton wrote the postscript in part to fend off Leibniz's and others' criticisms of his apparent invocation of action at a distance.

137 **"I have not as yet been able":** Newton, *Principia*, 943.

137 **frees scientific theory builders to try:** Many historians have remarked on Newton's successful attempt to move the explanatory goalposts, offering varying interpretations of the resulting framework. I. Bernard Cohen, for example, distinguishes a "Newtonian style" (in fact, a thoroughgoing method) in which explanation is shallow in my sense but in addition explicitly mathematical (Cohen, *Newtonian Revolution*).

140 **each is warmly welcomed:** Scientists may, of course, assign different plausibility rankings to different kinds of explanations. The iron rule does not demand that scientists treat all explanatory principles equally. But it does demand that they recognize any shallow causal principle's right to participate in scientific argument as a potential explanation.

There are two important differences between denying a principle this right and merely assigning it a low plausibility ranking. The first is that assigning a low ranking makes the evaluation of explanations in science a matter of degree rather than an exercise in censorship; consequently, even the most dimly regarded hypothesis can, if it achieves sufficient explanatory success, come to dominate the discussion. The second is that assigning rankings brings disputes as to which explanatory principle is best inside the framework of science, where they can be resolved using the empirical power of the knowledge machine.

140 **By the time Newton died:** The quoted tributes can be found in Mordechai Feingold, *Newtonian Moment*, 169, 173, 177.

142 **"science personified":** Feingold, *Newtonian Moment*, 148.

143 **"Elementary material particles":** Quoted in Helge Kragh, *Quantum Generations*, 117. Kragh gives a fascinating account of what he calls the "electromagnetic worldview."

145 **"Bohr from out of philosophical smoke clouds":** Quoted in Kragh, *Quantum Generations*, 213.

145 **"Quantum mechanics is certainly imposing":** In Born and Einstein, *The Born-Einstein Letters*, 91.

146 **"met almost no resistance":** According to Kragh, *Quantum Generations*, 169. On the marginalization of the philosophical aspects of quantum theory in the wider scientific world, see *Quantum Generations*, 211.

146 **"mysterious, confusing discipline":** These and many other such remarks can be found at https://en.wikiquote.org/wiki/Quantum_mechanics.

146 **let me take you wading:** Here I describe the "non-relativistic" version of the theory that was the subject of the philosophical debates of the 1920s and that is the form of quantum mechanics you would first encounter as an undergraduate physics major today.

147 **All that concerns the iron rule:** Because the iron rule looks for the purpose of testing to the observable consequences of causal principles, some thinkers have made the mistake of supposing that science cares only about things that can be directly observed. That is clearly not the case: astronomers care about the interiors of stars, biologists care about the molecular makeup of genomes, and physicists care about the architecture of atoms. Insofar as these unseen entities have divergent causal consequences, their structure can be inferred using the iron rule.

147 **the wave function is a complete description:** I am notionally including a specification of the electron's "spin" in the wave function. Experts will appreciate that the exclusive focus on position in my description of the quantum machinery is a simplification; it is not, I think, a distortion.

148 **only three dimensions:** In some versions of what physicists call string theory, the universe has more than three dimensions: 10, 11, or even 26. But these dimensions are "rolled up"; for this and other reasons, they are not suitable for the accommodation of the kind of ether that could provide a substrate for quantum waves. They have other work to do.

149 **To sum up, quantum mechanical matter:** This summary skates over numerous subtle interpretative issues. Does an observed particle really acquire a definite position, or does it just act that way? On some understandings of quantum mechanics (the "many worlds" and "many minds" interpretations) it is the latter. Is the wave function really a complete description of a particle's state? On one understanding of quantum mechanics (Bohmian mechanics), no: particles have definite positions at all times, even when they are exhibiting wavelike behavior. For an accessible and sophisticated introduction to these questions, try David Albert's *Quantum Mechanics and Experience*.

151 **"When confronted with a choice":** Heilbron, *Electricity in the 17th and 18th Centuries*, 500.

CHAPTER 7: THE DRIVE FOR OBJECTIVITY

152 **Over the walls of a moonlit Spanish cemetery:** As described in Santiago Ramón y Cajal, *Recollections of My Life*, 144–145. The quote is from p. 144.

154 **"admirably convenient, since it did away":** Cajal, *Recollections of My Life*, 336.

154 **"artificially distorted and falsified":** Quoted by by Lorraine Daston and Peter Galison in their presentation of the case in *Objectivity*, 116.

154 **"Objectivity was at once":** Daston and Galison, *Objectivity*, 120.

154 **"Thirsting for the objective and the concrete":** Cajal, *Recollections of My Life*, 145.

155 **"exactly prepared according to nature":** Quoted by Daston and Galison, *Objectivity*, 116.

159 **"Thus the results of the expeditions":** Dyson, Eddington, and Davidson, "Determination of the Deflection of Light," 332.

159 **"diffused and apparently out of focus":** Dyson, Eddington, and Davidson, "Determination of the Deflection of Light," 309.

159 **"much less weight":** Dyson, Eddington, and Davidson, "Determination of the Deflection of Light," 312.

160 **"difficult to say":** Dyson, Eddington, and Davidson, "Determination of the Deflection of Light," 309.

160 **He does nothing of the sort:** Also missing from the paper is an explanation of why the 4-inch telescope's mirror would not have been affected in the same way as the astrographic's mirror by the sun's heat.

161 **arguments appearing in official scientific venues:** Sometimes the conclusion is weakened to reflect the thinness of the argument: it is not unusual for a paper to conclude with nothing more than the rather anodyne comment that the observed data is "consistent with" the hypothesis under scrutiny.

162 **"There is hardly any difference":** Feyerabend, *Science in a Free Society*, 82.

164 **the singular kind of objectivity achieved by sterilization:** There is a parallel between science's focus on objectivity in specific official venues and the institution of courts of law, in which arguments in the courtroom are also governed by stringent rules dictating what kinds of considerations may and may not appear. The courts' restrictions, however, stem more from principles of fairness, such as the US Constitution's prescription that "in all criminal prosecutions, the accused shall enjoy the right . . . to be confronted with the witnesses against him," than from a concern with objectivity for its own sake.

166 **"vast pile":** Sprat, *History of the Royal Society of London*, 118.

166 **"The Society has reduced its principal observations":** Sprat, *History of the Royal Society of London*, 115.

167 **"We took a slender and very curiously blown cylinder":** Quoted in Dear, "Totius in Verba," 153. Dear's own acute characterization of the passage, and of the house style of the early *Transactions* in general, is from the same page.

168 **can be gamed to illuminate the data:** The process of running through many different statistical analyses to squeeze some significant result from a set of observations is called "data dredging" or (referring to a particular class of statistical methods) "*p*-hacking." You can try out *p*-hacking yourself at http://fivethirtyeight.com/features/science-isnt-broken/ (Aschwanden, "Science Isn't Broken").

169 "The particular and endless modifications": Scoresby, *Account of the Arctic Regions*, vol. 1, 426–7.

169 During one particularly brutal winter freeze: As described in Glaisher, "On the Severe Weather at the Beginning of the Year 1855," 16–30.

171 "the old-school version of Photoshop": Libbrecht's comparison to Photoshop and his snowflake statistics are from Pilcher, "No Great Flakes," 71.

172 even the camera has a point of view: Many further fascinating aspects of the changing notion of objectivity in scientific imagery, including both the Cajal-Golgi dispute and the case of snowflakes, are described in Daston and Galison's book *Objectivity* and in the literature on which it draws.

CHAPTER 8: THE SUPREMACY OF OBSERVATION

173 "Experiment is the sole judge of scientific truth": From the opening pages of Feynman's *Lectures on Physics*, Herschel's *Preliminary Discourse on the Study of Natural Philosophy* (p. 80), Medawar's *Induction and Intuition in Scientific Thought* (p. 42), and Hawking and Roger Penrose's *Nature of Space and Time* (p. 121).

174 Whewell's "temper will never be good!": Quoted in Laura Snyder, *Philosophical Breakfast Club*, 209. Snyder's book provides an excellent introduction to Whewell's life and accomplishments.

174 "In early days": From Leslie Stephen's entry on Whewell in *The Dictionary of National Biography*, 460.

176 "The species of plants and animals": Whewell, *History of the Inductive Sciences*, vol. III, 569.

176 "other powers": Whewell, *History of the Inductive Sciences*, vol. III, 582.

177 would not emerge for another 20 years: Arriving sooner, in 1844, would be the Scottish journalist Robert Chambers's anonymously published *Vestiges of the Natural History of Creation*, a book that told an evolutionary story in which simpler organisms were transmuted into more complex organisms, ending with the evolution of the human species. Unlike Darwin, Chambers attributed it all to a Creator: evolution was in accordance with "a natural principle flowing from [God's] mind" (p. 154). He was in this respect a man of his time, far closer in his thinking to Whewell than to Darwin.

177 "every advance in our knowledge": From Whewell's Bridgewater Treatise, *Astronomy and General Physics Considered with Reference to Natural Theology*, vi, 252–3.

178 "must never be allowed": Whewell, *History of the Inductive Sciences*, vol. III, 583.

178 "the ordinary evidence of science": Whewell, *History of the Inductive Sciences*, vol. III, 582.

178 theology was able to fill these blanks: Whewell did prevaricate a little, maintaining that on theological grounds alone it would be difficult to predict "the detail of all events in the history of man, or of the skies, or of the earth," or at least that the attempt to do so would not be "immune from error" (*History of the Inductive Sciences*, vol. III, 585). But this proviso, to someone with Whewell's beliefs, hardly precluded theology's having something relevant and interesting to say about the creation of species: even if it did not supply "the detail of all events," it might have contributed considerably to humanity's understanding of their broad outlines.

179 He repudiated the project: Thomas Burnet's *Sacred Theory of the Earth*, published in the late seventeenth century, gives a sense of what such a project might look like.

Burnet's subject matter is, like Whewell's, the geological history of the earth. His argument effortlessly and cogently mingles physics and scriptural interpretation, both his own and that of earlier Christian thinkers. The method looks strange to intellects like ours, habituated to the ways of modern science, but it makes perfect sense if you suppose, as Burnet did, that sacred texts tell us something about the history of the world. Burnet was not a literalist; he uses common sense, his knowledge of natural processes such as the flow of rivers, and biblical scholarship to temper his application of the texts (see Figure 3.3). A presentation and appreciation of Burnet's thinking may be found in Stephen Jay Gould's *Time's Arrow; Time's Cycle.*

179 **"You have no idea"**: Quoted in Snyder, *Philosophical Breakfast Club*, 367–8.

182 **"that reason, whether finite or infinite"**: Whewell, *History of the Inductive Sciences*, vol. III, 586.

183 **"Concerning Magnesia or the Green Lion"**: The excerpts from Newton's notebooks in this chapter are taken from Richard Westfall's biography of Newton, *Never at Rest*, 292, 367–8.

185 **"well over a million words"**: According to Westfall, *Never at Rest*, 290.

185 **"to liberate the spirit or active virtue"**: An interpretation suggested by Westfall, *Never at Rest*, 364.

186 **"The vital agent"**: Quoted in Westfall, *Never at Rest*, 304.

187 **a belief in the *prisca sapientia***: Most notable among those emphasizing Newton's commitment to the *prisca sapientia* are J. E. McGuire and P. M. Rattansi ("Newton and the 'Pipes of Pan'"), Betty Jo Dobbs, who published the first full-length scholarly study of Newton's alchemy in 1975 (*Foundations of Newton's Alchemy*), and Frank Manuel (*Religion of Isaac Newton*).

187 **"the net"**: Some of Newton's references to the legend are compiled by Westfall, *Never at Rest*, 296.

187 **Francis Bacon, decades earlier, warned**: In his *New Organon*, §1.65.

188 **"the last of the magicians"**: Keynes, "Newton, the Man," 27.

188 **His intellect operated**: There are no absolutes in psychology, not even Newton's; thus, it is possible to find incipient signs of integration in his work. He attempted to fuse his alchemy with the atomist conception of matter; his physics borrows the idea of absolute space from his philosophy (although with an additional empirical justification); his theory of gases borrows particles from his work on light and borrows the forces between them (albeit repulsive) from his work on gravity.

Some historians have made a case for hidden connections between Newton's various inquiries. David Castillejo used numerological arguments to claim that Newton's thought was "ruthlessly interlocked"; he saw a correspondence, for example, between the four sides of Newtonian light rays and the four walls of another of Newton's interests, Solomon's temple. But contemporary historians have by and large conceded the obvious: "There is surprisingly little cross-referencing of themes from one area of Newton's endeavors to another," writes George Smith ("Isaac Newton"). Sarah Dry offers an overview of recent Newton interpretation in *The Newton Papers* (on Castillejo, see pp. 200–201).

192 **Newton gave his successors**: Reading Newton's prescription carefully, you can see in addition the germ of the iron rule's other negative injunction, its prohibition on anything subjective. Hypotheses, Newton holds, must be established by *deducing* them from the phenomena, a formulation that suggests an appeal to objective laws

of logic alone. In a few words, then, Newton encapsulates and entrusts to posterity the entirety of the iron rule's negative side.

For the same reason that contemporary philosophers of science believe that scientific reasoning is essentially subjective, they doubt that Newton or any other scientist has ever successfully deduced a hypothesis from observable phenomena alone: plausibility rankings are always part of the mix. Newton's formulation therefore serves more as an exhortation toward objectivity than as an accurate summary of his method. As I concluded in Chapter 7, however, an attempt to make scientific argument objective—to "sterilize" the subjectivity away—can be highly effective even when not wholly successful.

192 **"It is no less of a problem"**: Bacon, *New Organon*, §66.

193 **Bacon's rationale for shallow explanation**: Bacon is famous for remarking (without quite using this pithy expression) that knowledge is power. In making the case for shallow explanation, he seems to advocate the coarser doctrine that knowledge is valuable only insofar as it is power. Perhaps this is what William Harvey, who discovered the circulation of blood and was for a time Bacon's physician, was insinuating when he said that Bacon "writes philosophy like a Lord Chancellor."

193 **"To keep my judgement"**: From the preface to Boyle's *Some Specimens of an Attempt to Make Chymical Experiments Useful*, 355. The preface was most likely written around 1660.

194 **modern historians doubt**: Boas, *Robert Boyle and Seventeenth-Century Chemistry*, 26–28.

194 **He needed no direction**: We know that Newton read Boyle, but as far as we know he never read Bacon. Nor may he have read Galileo's most Newtonian work, the *Two New Sciences* (Westfall, *Never at Rest*, 89; Cohen "A Guide to Newton's *Principia*," in Cohen and Whitman's translation of the *Principia*, 146).

CHAPTER 9: SCIENCE'S STRATEGIC IRRATIONALITY

201 **"Credit must be given"**: Aristotle, *On the Generation of Animals*, III.10, 760b29–30, translated by Arthur Platt, in *Complete Works of Aristotle*.

201 **It makes for a fascinating story**: Readers interested in learning more about the causes of the Scientific Revolution will enjoy H. Floris Cohen's *Rise of Modern Science Explained* or David Wootton's *Invention of Science*. Cohen and Wootton ask, like me, "What exactly was it that enabled seventeenth- and eighteenth-century science to make progress in a way that previous systems of knowledge could not?" (Wootton's words, p. 4); unlike me, they present their answers in the form of histories of the origins of modern science that discuss the role of a wide variety of historical factors. For an argument that the Scientific Revolution happened when it did as a matter of chance, read J. D. Trout's provocative *Wondrous Truths*.

204 **"Their explanation of the phenomena"**: Aristotle, *On the Heavens*, III.7, 306a, translated by J. L. Stocks, in *Complete Works of Aristotle*.

205 **Augustine, Avicenna, Averroes, and Aquinas**: Augustine was a fourth-century Roman theologian and philosopher; Avicenna (980–1071) and Averroes (1126–1198) were Islamic philosophers working in Persia and Spain, respectively; Aquinas (1225–1274) was an Italian theologian, philosopher, and monk.

207 **"The flood-gates of infidelity"**: In Romanes's anonymously published *Candid Examination of Theism*, 51–52.

207 **perhaps a third of American scientists believe in God:** This statistic is drawn from a survey by the Pew Research Center for the People & the Press conducted in 2009 (https://www.pewforum.org/2009/11/05/scientists-and-belief/). According to the same survey, about 40 percent of scientists say they do not believe in God or a higher power. These numbers seem not to have changed much in 100 years.

207 **"non-overlapping magisteria":** Gould writes about this notion in *Rocks of Ages*.

207 **"Shut up and calculate":** In the rather conflicted words of the thoughtful physicist David Mermin, "What Is Wrong with This Pillow?," 9.

208 **"I know of *no one*":** Weinberg, *Dreams of a Final Theory*, 168–9. Is philosophy really so useless to science? In fact, it has plenty to offer scientists thinking about the meaning of quantum mechanics, as Weinberg sometimes does himself (https://www.nybooks .com/articles/2017/01/19/trouble-with-quantummechanics/). But there is no doubt, as I said at the beginning of this chapter, that philosophical reasoning has turned out to be far less useful in empirical inquiry than natural philosophers such as Aristotle once supposed. The reason, above all, is the Tychonic nature of our world, which offers clues that must be won not by deep thought, but by unstinting observation.

CHAPTER 10: THE WAR AGAINST BEAUTY

212 **"He looked on the whole universe":** Keynes, "Newton, the Man," 29.

215 **the classification of birds:** The diagrams in Figures 10.4 and 10.5 are based on those in Vigors, "Observations on the Natural Affinities," 468, 509.

216 **A group shared "affinities":** As well as affinities, according to Macleay, groups were related by "analogies" that connected groups occupying the same position, relative to the center, on different circles.

216 **new taxa with the corresponding affinities:** In this connection I can't resist mentioning Aristotle's attempt to use the rule of four to predict the existence of as yet unobserved creatures. He reasons that three broad classes of animals correspond to three of the four elements: "Plants may be assigned to the land [that is, earth], the aquatic animals to water, the land animals to air." There must be a fourth class of animals, then, that corresponds to the fourth element, fire. Why are they nowhere to be found? In a stroke of exquisite biological lunacy, Aristotle suggests that they are living on the moon. (*On the Generation of Animals*, III.11, 761b, translated by Arthur Platt, in *Complete Works of Aristotle*.)

217 **life has the grand design:** The nested quinarian structure is not a true fractal, because the embedding does not go on forever but rather bottoms out at the species level—although Macleay came close to imagining a continuum of species in which infinite nesting might be possible, writing, "If we knew *all* the species of the creation, their number would be infinite, or in other words, . . . they would pass into each other by infinitely small differences" (*Illustrations of the Annulosa of South Africa*, 8n).

217 **"Nature appeared to me to have branched out":** Macleay, *Horae Entomologicae*, 170.

218 **One such follower:** Darwin's flirtation with quinarianism is described in Dov Ospovat, *Development of Darwin's Theory*, 101–13.

219 **"to study Nature simply as she exists":** Strickland, "On the True Method of Discovering the Natural System in Zoology and Botany," 192.

221 **"I know that in the study of material things":** Thompson, *On Growth and Form*, 326.

222 "It accounts, by one single integral transformation": Thompson, *On Growth and Form*, 300.

223 "a comprehensive 'law of growth'": Thompson, *On Growth and Form*, 275.

226 his high regard for beauty and form: In touting the importance of beauty, did Thompson violate the iron rule? Like many of the most creative scientific thinkers, he bent it to near breaking point but then relented and followed its prescription. A distinction must be made between a scientist's acknowledging the role of beauty in the generation of their ideas, on the one hand, and using considerations of beauty to argue for their ideas, on the other. Thompson's rhetoric comes close to the latter, which would contravene the iron rule, but a close inspection of his argument reveals that it turns principally on the ability of certain hypotheses to explain certain observations—thus, the "law of growth" is "proved" by its ability to explain the simple geometrical relations that hold between body plans. Thompson certainly hopes that the reader will see the aesthetic appeal of his ideas—no question of that!—but like every scientist laboring under the iron rule's dominion, he knows that he ought to show rather than to tell.

227 "It is more important to have beauty": In Dirac's 1963 essay "Evolution of the Physicist's Picture of Nature," p. 47. Dirac used Schrödinger's Nobel Prize–winning work on quantum mechanics as an example of his precept; he might equally well have cited his own.

227 "The only physical theories which we are willing to accept": This famous formulation appears in Wigner's equally famous paper "The Unreasonable Effectiveness of Mathematics in the Natural Sciences." I haven't been able to find an independent source for the words Wigner attributes to Einstein; he may have intended them only as a paraphrase of Einstein's well-known regard for a theory's aesthetic merits.

227 You needn't look far to find similar sentiments: Chandrasekhar, *Truth and Beauty*; Deutsch, *Beginning of Infinity*, Chapter 14; Wilczek, *Beautiful Question*. Sabine Hossenfelder voices a dissenting view in *Lost in Math*.

232 two new xi particles: Note the additional star that distinguishes these xi particles from those in the octet pictured in Figure 10.11. The difference is in the particles' "spin": the starred xis have spin 3/2, whereas the unstarred xis have spin 1/2.

233 It was exactly what the iron rule demanded: Gell-Mann's scheme also implied the existence of another particle, the eta, just as experimenters were beginning to see evidence for it in their data.

233 Murray Gell-Mann had secured his Nobel Prize: The prize was awarded in 1969. Along with the Nobel's exclusivity come invidious distinctions. Each of Gell-Mann's great breakthroughs had been made independently and at the same time by another scientist: Kazuhiko Nishijima also discovered strangeness; Yuval Ne'eman, the eightfold way; George Zweig, quarks. No one, however, would dispute that Gell-Mann was a deserving winner.

234 "a chief criterion for the selection of the correct hypothesis": Quoted in George Johnson's biography of Gell-Mann, *Strange Beauty*, 239.

234 Yet he made no appeal: Gell-Mann's 1962 technical paper on the eightfold way characterizes the scheme in passing as "appealing," but the appearance of that one word is as close as Gell-Mann comes to making an aesthetic case for his theory. He spends several pages discussing possible empirical tests of his ideas.

235 "make choices and exercise judgments": B. Greene, *Elegant Universe*, 166–7.

235 **"aesthetic judgments do not arbitrate scientific discourse":** B. Greene, *Elegant Universe*, 166.

236 **a scorebook should be maintained:** This is the advice offered by the philosopher of science James McAllister in his book *Beauty and Revolution in Science.*

237 **That is illogical, unreasonable, irrational:** Technically, what the iron rule does wrong from a logical point of view is to violate what philosophers call the "principle of total evidence," a rule of rationality that requires that in making a case for a view, you must (insofar as is practically possible) take every relevant consideration into account. In scientific reasoning, aesthetic reasons are relevant, but the iron rule ignores them.

I must emphasize that the iron rule does not force individuals to contravene the principle of total evidence in their private scientific reasoning. Many take full advantage of this latitude, incorporating into their evaluation of theories whatever kinds of nonempirical reasons—aesthetic, philosophical, religious—they take to be important. Indeed, some of the most thoughtful people I know are psychologists, physicists, linguists, biologists. It is not scientists' thinking, but science as a social process of cooperative inquiry, as a mode of intellectual engagement, as a set of rules for conducting a public argument, that disregards the rules of logic.

CHAPTER 11: THE ADVENT OF SCIENCE

242 **"I realized that it was necessary":** Descartes, *Meditations* I (published in 1641).

243 **"A new beginning has to be made":** Bacon, *New Organon*, §1.31; also quoted in Chapter 5.

243 **without precedent in human history:** Fifth- and fourth-century BCE Athens is perhaps the seventeenth century's only serious rival.

243 **such tremendous variety:** Many historians have attempted to explain the early modern European penchant for novelty. Was it enabled by the world's first mass medium, the newly invented movable-type printing press? Was it inspired by the discovery of the New World and the growth of international trade? Was it swept along by the economic and demographic recovery from the devastation of the Black Death? I won't take a stand on this question; it's enough that among many other things it happened to throw up a rather thoroughgoing empiricism.

243 **the explanation has barely begun:** To suppose that the "right" ideas, once someone goes to the trouble to have them, will inevitably triumph over the rest, as though they are taken up on the shoulders of a band of heroes and borne toward the future along a rose-petaled path, is to indulge in the fallacy that the scholar Herbert Butterfield termed "whig history."

244 **Boyle wrote a great deal:** Boas, *Robert Boyle and Seventeenth-Century Chemistry*, 75–78. "little bodies": quoted on p. 82.

246 **"The terms Protestant and Catholic":** Wedgwood, *Thirty Years War*, 373.

246 **"It was not that faith":** Wedgwood, *Thirty Years War*, 372.

246 **For civic purposes, what mattered:** Although Britain was not directly involved in the Thirty Years' War, much the same effect was realized by the English Civil War of 1642–1651 and its aftermath.

247 **separation of political and religious identity:** Louis XIV attempted, for example, to reimpose Catholicism on all of France in 1685.

247 "The laws of God & the laws of man": Newton, "Seven Statements on Religion."

247 a permanent bisection: The German political philosopher and jurist Samuel von Pufendorf, born among the battlefields of the Thirty Years' War in Saxony in 1632, wrote memorably of an individual's different "moral personae"—as citizen, shopkeeper, spouse—each governed by a distinct code. For the sake of peace, it was imperative that these codes not conflict. They must be confined to separate areas of life, or, where their jurisdictions overlapped, there must be a clear, universally agreed on order of precedence among them. (Pufendorf's moral agents included not only individuals but also associations of various kinds; religions entered into his system as corporate bodies operating within and across states.) In the seventeenth century, Pufendorf's influence was immense; then, as his thought lost its pertinence in an increasingly liberal Europe, his reputation waned.

250 fraudulent emendations: In the fourth century, the Church Father Athanasius led a campaign to establish the doctrine of the trinity as orthodoxy within the Catholic Church. Newton conjectured that certain parts of the New Testament—in particular, the first letter of John and the letter to Timothy—were later deliberately altered to favor the doctrine, for which he discerned no support in the earliest Christian writings.

250 "branded a moral leper": In the words of Newton's biographer Richard Westfall. The story is related in Westfall's *Never at Rest*, 330–34.

251 a conflict that arose repeatedly: The most famous case is that of Galileo, convicted of heresy by the Catholic Church in 1633. Galileo did his best to maintain that he neither held nor expressed the view that the earth orbited the sun, nor any other proscribed opinions, but the Inquisition was not ready to make compromises of the sort that later saved Newton's career. Rather than taking Galileo at his word, the inquisitors insisted on looking into his heart—where they discerned, quite accurately, his true beliefs.

251 "Do not befriend kings": Al-Nuwayrī, *Ultimate Ambition in the Arts of Erudition*, 96.

251 The alternative—perpetual religious conflict: According to the seventeenth century's great genius of political philosophy, Thomas Hobbes, the most common cause of civil war is the difficulty of "obeying at once both God and Man." War has to be avoided at all costs; consequently, Hobbes maintains, even when the monarch is, from the subject's point of view, an "infidel," the subject must obey the law—if necessary conforming publicly to the rites of a state religion that they consider to be corrupt or evil. God, he continues somewhat optimistically, cares only for what lies on the inside: "And for their faith, it is internal and invisible; they . . . need not put themselves into danger for it." (Quotes from Hobbes, *Leviathan*, III.43.)

CHAPTER 12: BUILDING THE SCIENTIFIC MIND

255 "The average person cannot imagine": Jahren, *Lab Girl*, 54.

255 "Every feature of every hole": Jahren, *Lab Girl*, 54–55.

256 "indefatigable persistence": Cajal, *Recollections of My Life*, 278.

256 "We would not accept": Weinberg, *Dreams of a Final Theory*, 165.

258 one of Thomas Kuhn's more controversial papers: Kuhn, in "Function of Dogma in Scientific Research," argues that it is on the whole advantageous to teach the doctrines of the prevailing paradigm to young scientists uncritically.

260 "Philosophy is dead": Hawking and Mlodinow, *Grand Design*, 5.

261 **"My concern here"**: The interview with Tyson is available at https://id10t.com/ podcast/episode-489-neil-degrasse-tyson-returns-again/.

261 **"Philosophy is a field"**: Krauss's interview can be read at http://www.theatlantic .com/technology/archive/2012/04/has-physics-made-philosophy-and-religion -obsolete/256203/.

261 **"It's shocking that such brilliant scientists"**: Olivia Goldhill, "Why Are So Many Smart People Such Idiots about Philosophy?," *Quartz*, http://qz.com/627989/why -are-so-many-smart-people-such-idiots-about-philosophy/.

262 **"Philosophers' historic failure"**: Dawkins's tweet can be read at https://twitter .com/richarddawkins/status/433519270102708224.

263 **"Would a leading North American journal"**: Quotes are from Alan Sokal, "A Physicist Experiments with Cultural Studies," originally published in the journal *Lingua Franca*. This article and many related resources can be found on Sokal's website at https://physics.nyu.edu/faculty/sokal/.

263 **"Tell most scientists about the 'science wars'"**: Gould, *The Hedgehog, the Fox, and the Magister's Pox*, 101–2.

264 **"So many scientists are narrow, foolish people"**: Wilson, *Consilience*, 62; it is narrowness rather than foolishness that I emphasize here. For similar thoughts, see *Consilience*, 42, and Ziman, *Real Science*, 161.

264 **"A good commander"**: Tolstoy, *War and Peace*, 644.

264 **It seems apt**: Schally's comparison is recorded by Latour and Woolgar, *Laboratory Life*, 130.

265 **These thinkers**: I hope I have not omitted any friends or important reviewers.

267 **"Resolve to abide"**: Lawrence, *Studies in Classic American Literature*, 27.

267 **"To be nobody-but-yourself"**: Cummings, "A Poet's Advice to Students," 13.

267 **"In this ever-changing society"**: Schultz and Yang, *Pour Your Heart Into It*, 248.

CHAPTER 13: SCIENCE AND HUMANISM

269 **"Let us be driven"**: Pico della Mirandola, *Oration on the Dignity of Man*, 26.

272 **what is found beyond the golden gate**: The opposition between the two gates stands at the heart of Proust's great novel *In Search of Lost Time*, from which I have stolen these words (taken from the second volume, adapting the Moncrieff/Kilmartin/ Enright translation, p. 377).

274 **"the standard schoolbook"**: According to H. O. Taylor, *Classical Heritage of the Middle Ages*, 50. Says John Sandys: "In the earlier Middle Ages it was the principal, often the only, textbook used in schools, and it exercised a considerable influence on education and on literary taste" (*Short History of Classical Scholarship*, 68).

274 **The bridegroom Mercury stands for eloquence**: The interpretation offered here— with Mercury representing the three humanistic liberal arts and Philology the four scientific arts—was standard throughout the Middle Ages and is also thought by modern scholars to be what Martianus intended.

275 **"If the heuristic and analytic power of science"**: Wilson, *Meaning of Human Existence*, 187. Wilson asserts two pages earlier that it is an "archaic misconception" to suppose that "the farther apart [the sciences and the humanities] are kept, the better" (p. 185).

CHAPTER 14: CARE AND MAINTENANCE
OF THE KNOWLEDGE MACHINE

279 **rising sea levels may have forced:** According to Kulp and Strauss, "New Elevation Data Triple Estimates of Global Vulnerability."

280 **the question of agenda setting:** To set science's priorities sensibly, we must balance three demands. First, we need science to address our most pressing needs. That said, we must not neglect more fundamental research that will provide a basis for attending to future needs whose nature is as yet only dimly perceived. And finally, even pure knowledge has its value; we shouldn't neglect research whose insights into the workings of the world will make our lives, assuming that we manage to save them, intellectually richer.

Some writers have suggested that citizen committees should have a say in determining the direction of scientific research: Sheila Jasanoff, in "Transparency in Public Science," and Stephen Turner, in *Liberal Democracy 3.0*, are two. The philosopher Philip Kitcher, in *Science in a Democratic Society*, is also favorably disposed, at least in principle. I myself believe that science's internal system of incentives renders it far more sensitive to society's priorities than might be supposed. It could almost be left to make up its own mind about what matters. But as I have said, that topic must wait until another time.

282 **"investigate some part of nature":** Kuhn, *Structure of Scientific Revolutions*, 25, cited earlier in Chapter 5.

282 **A Kuhnian program of science education:** As outlined in Kuhn's "Function of Dogma in Scientific Research, cited in Chapter 12.

284 **detectors the size of the planet Jupiter:** These are back-of-the-envelope estimates mentioned by many commentators, for example, Sabine Hossenfelder in *Lost in Math*, 178. For a more systematic exposition of the obstacles to testing string theory, you might look at Lee Smolin's *Trouble with Physics*.

284 **One suggested revision:** Physicists have not on the whole phrased their amendments in terms of what is allowed in the "official channels of scientific communication," but that seems to me to be the best interpretation of what they have in mind. The philosopher Richard Dawid makes a case along these lines—invoking the idea, mooted at the end of Chapter 10, that we can learn which kinds of explanatory elegance are signs of truth—in *String Theory and the Scientific Method*.

284 **"post-empirical physics":** Another amendment to the iron rule would allow string theory to accrue scientific legitimacy by providing explanations of the values of the physical constants—which the Standard Model treats as brute posits—in a way that departs from the canon of shallow causal explanation, by appeal to what is called an "anthropic principle." Anthropic explanations invoke the existence, in our universe, of intelligent observers, remarking that we should expect to see only physical constants that would allow complex life-forms like ourselves to evolve. Scientists and philosophers argue about whether such an aperçu genuinely helps to account for the physical constants having the (approximate) values that they do. What's clear is that anthropic derivations are not explanations of the shallow causal sort allowed by the iron rule: they invoke a fact, the presence of watchful intellects, that is causally irrelevant to the values of the constants. Thus, even if string theory, together with the anthropic principle, gives us reason to expect certain values, that

prediction or explanation (if it is such) does not, by the lights of the iron rule, constitute the basis for a legitimate empirical test.

285 **"It will follow orders"**: Collins and Pinch, *The Golem*, 1.

286 **"bumbling giant"**: Collins and Pinch, *The Golem*, 2.

288 **they assign confidence levels to hypotheses**: These terms have precisely specified meanings. A hypothesis endorsed with very high confidence is supposed to have a 9 out of 10 chance of proving correct. A very likely event is supposed to occur with a probability of between 90 and 95 percent. The two scales are combined; thus, the IPCC can state "with high confidence" that a certain event is "more likely than not."

288 **the IPCC aims to use a range of rankings**: The IPCC's "medium confidence" and "more likely than not," then, do not entirely reflect the variation in opinion inherent in the full range of plausibility rankings across the scientific community. "Medium confidence," for example, is compatible with the existence of a substantial minority of scientists who vehemently disagree. An assessment that reflected the complete span of scientific opinion would on many important questions be so indeterminate as to be practically useless.

289 **"if we care about the future"**: In Schneider's illuminating discussion of the IPCC's efforts to codify subjectivity, "Confidence, Consensus and the Uncertainty Cops," 434.

290 **"I had always felt so unable"**: In a letter from Galileo to Fortunio Liceti, quoted in Frova and Marenzana, *Thus Spoke Galileo*, 414.

REFERENCES

Albert, D. Z. *Quantum Mechanics and Experience*. Cambridge, MA: Harvard University Press, 1992.

Al-Nuwayrī, S. *The Ultimate Ambition in the Arts of Erudition*. Translated by E. Muhanna. New York: Penguin, 2016.

Aristotle. *The Complete Works of Aristotle: The Revised Oxford Translation*. Edited by J. Barnes. Princeton, NJ: Princeton University Press, 1984.

Aronowitz, S. *Science as Power: Discourse and Ideology in Modern Society*. Minneapolis: University of Minnesota Press, 1988.

Bacon, F. *The New Organon*. Translated by M. Silverthorne. Edited by L. Jardine and M. Silverthorne. Cambridge: Cambridge University Press, 2000.

Barnes, D. E., and L. A. Bero. "Why Review Articles on the Health Effects of Passive Smoking Reach Different Conclusions." *Journal of the American Medical Association* 279 (1998): 1566–70.

Barnes, J. *Aristotle*. Oxford: Oxford University Press, 1982.

Bekelman, J. E., Y. Li, and C. P. Gross. "Scope and Impact of Financial Conflicts of Interest in Biomedical Research: A Systematic Review." *Journal of the American Medical Association* 289 (2003): 454–65.

Bes-Rastrollo, M., M. B. Schulze, M. Ruiz-Canela, and M. A. Martinez-Gonzalez. "Financial Conflicts of Interest and Reporting Bias Regarding the Association between Sugar-Sweetened Beverages and Weight Gain: A Systematic Review of Systematic Reviews." *PLoS Medicine* 10, no. 12 (2013): e1001578.

Biagioli, M. "Productive Illusions: Kuhn's *Structure* as a Recruitment Tool." *Historical Studies in the Natural Sciences* 42 (2012): 479–84.

Boas, M. *Robert Boyle and Seventeenth-Century Chemistry*. Cambridge, UK: Cambridge University Press, 1958.

Born, M., H. Born, and A. Einstein. *The Born-Einstein Letters: Correspondence between Albert Einstein and Max and Hedwig Born from 1916 to 1955 with Commentaries by Max Born.* Translated by I. Born. London: Macmillan, 1971.

Boyle, R. *Some Specimens of an Attempt to Make Chymical Experiments Useful to Illustrate the Notions of the Corpuscular Philosophy.* In *The Works of the Honourable Robert Boyle in Six Volumes.* Vol. 1. Edited by T. Birch, 354–9. London: J & F Rivington, 1772.

Burchfield, J. D. *Lord Kelvin and the Age of the Earth.* New York: Science History Publications, 1975.

Burnet, T. *The Sacred Theory of the Earth.* Vol. 1. London: R. Norton, 1684.

Campbell, J. *The Lives of the Lord Chancellors and Keepers of the Great Seal of England.* Vol. 2, 3rd ed. London: John Murray, 1848.

Chambers, R. *Vestiges of the Natural History of Creation.* London: John Churchill, 1844.

Cajal, S. Ramón y. *Recollections of My Life.* Translated by E. H. Craigie. Philadelphia: American Philosophical Society, 1937.

Chandrasekhar, S. *Truth and Beauty: Aesthetics and Motivations in Science.* Chicago: University of Chicago Press, 1987.

Cho, M. K., and L. A. Bero. "The Quality of Drug Studies Published in Symposium Proceedings." *Annals of Internal Medicine* 124 (1996): 485–9.

Christianson, G. E. *Isaac Newton.* Oxford: Oxford University Press, 2005.

Cohen, H. F. *The Rise of Modern Science Explained: A Comparative History.* Cambridge: Cambridge University Press, 2015.

Cohen, I. B. *The Newtonian Revolution: With Illustrations of the Transformation of Scientific Ideas.* Cambridge: Cambridge University Press, 1980.

Collins, H. M. "Stages in the Empirical Programme of Relativism." *Social Studies of Science* 11 (1981): 3–10.

Collins, H. M., and T. Pinch. *The Golem: What You Should Know about Science.* 2nd ed. Cambridge: Cambridge University Press, 2012.

Cummings, E. E. "A Poet's Advice to Students." In *E. E. Cummings: A Miscellany.* Rev. ed. Edited by G. J. Firmage. New York: October House, 1965.

Dalrymple, G. B. *The Age of the Earth.* Stanford, CA: Stanford University Press, 1991.

Darwin, C. *The Autobiography of Charles Darwin: With Original Omissions Restored.* Edited by N. Barlow. London: Collins, 1958.

Daston, L. J., and P. Galison. *Objectivity.* New York: Zone Books, 2007.

Dawid, R. *String Theory and the Scientific Method*. Cambridge: Cambridge University Press, 2013.

Dear, P. *The Intelligibility of Nature*. Chicago: University of Chicago Press, 2006.

Dear, P. "Totius in Verba: Rhetoric and Authority in the Early Royal Society." *Isis* 76 (1985): 145–61.

Debré, P. *Louis Pasteur*. Translated by E. Forster. Baltimore: Johns Hopkins University Press, 1998.

Descartes, R. *Meditations on First Philosophy*. In *The Philosophical Writings of Descartes*. Translated by J. Cottingham, R. Stoothoff, and D. Murdoch. Cambridge: Cambridge University Press, 1985.

Descartes, R. *Principles of Philosophy*. In *The Philosophical Writings of Descartes*. Translated by J. Cottingham, R. Stoothoff, and D. Murdoch. Cambridge: Cambridge University Press, 1985.

Descartes, R. *The World and Other Writings*. Translated by S. Gaukroger. Cambridge: Cambridge University Press, 1998.

Deutsch, D. *The Beginning of Infinity: Explanations That Transform the World*. New York: Viking, 2011.

Dirac, P. A. M. "The Evolution of the Physicist's Picture of Nature." *Scientific American* 208 (1963): 45–53.

Dobbs, B. J. T. *The Foundations of Newton's Alchemy, or "The Hunting of the Greene Lyon."* Cambridge: Cambridge University Press, 1975.

Dry, S. *The Newton Papers: The Strange and True Odyssey of Isaac Newton's Manuscripts*. Oxford: Oxford University Press, 2014.

Dyson, S. F. W., A. S. Eddington, and C. Davidson. "A Determination of the Deflection of Light by the Sun's Gravitational Field, from Observations made at the Total Eclipse of May 29, 1919." *Philosophical Transactions of the Royal Society of London, Series A* 220 (1920): 291–333.

Earman, J., and C. Glymour. "Relativity and Eclipses: The British Eclipse Expeditions of 1919 and Their Predecessors." *Historical Studies in the Physical Sciences* 11 (1980): 49–85.

Eccles, J. "Under the Spell of the Synapse." In *The Neurosciences: Paths of Discovery*, edited by F. G. Worden, J. P. Swazey, and G. Adelman. Cambridge, MA: MIT Press, 1975.

England, P. C., P. Molnar, and F. M. Richter. "Kelvin, Perry and the Age of the Earth." *American Scientist* 95 (2007): 342–9.

Fara, P. *Science: A Four Thousand Year History*. Oxford: Oxford University Press, 2010.

Feingold, M. *The Newtonian Moment: Isaac Newton and the Making of Modern Culture.* New York: New York Public Library and Oxford University Press, 2004.

Feyerabend, P. K. *Science in a Free Society.* London: New Left Books, 1978.

Feynman, R. P., R. B. Leighton, and M. Sands. *The Feynman Lectures on Physics.* Boston: Addison–Wesley, 1963.

Fisher, R. A. "Has Mendel's Work Been Rediscovered?" *Annals of Science* 1 (1936): 115–37.

Frova, A., and M. Marenzana. *Thus Spoke Galileo: The Great Scientist's Ideas and Their Relevance to the Present Day.* Translated by J. McManus. Oxford: Oxford University Press, 2006.

Garber, D. *Descartes' Metaphysical Physics.* Chicago: University of Chicago Press, 1992.

Gaukroger, S. *Descartes: An Intellectual Biography.* Oxford: Oxford University Press, 1995.

Giere, R. N. *Science without Laws.* Chicago: University of Chicago Press, 1999.

Glaisher, J. "On the Severe Weather at the Beginning of the Year 1855, and on Snow and Snow-Crystals." *Report of the Council of the British Meteorological Society*, 16–30, 1855.

Gleick, J. *Isaac Newton.* New York: Pantheon, 2003.

Gould, S. J. *The Hedgehog, the Fox, and the Magister's Pox: Mending the Gap between Science and the Humanities.* New York: Random House, 2003.

Gould, S. J. *Rocks of Ages: Science and Religion in the Fullness of Life.* New York: Ballantine Books, 1999.

Gould, S. J. *Time's Arrow; Time's Cycle: Myth and Metaphor in the Discovery of Geological Time.* Cambridge, MA: Harvard University Press, 1987.

Greene, B. *The Elegant Universe: Superstrings, Hidden Dimensions, and the Quest for the Ultimate Theory.* New York: W. W. Norton, 1999.

Greene, M. T. *Alfred Wegener: Science, Exploration, and the Theory of Continental Drift.* Baltimore: Johns Hopkins University Press, 2015.

Hawking, S., and L. Mlodinow. *The Grand Design.* New York: Bantam, 2010.

Hawking, S., and R. Penrose. *The Nature of Space and Time.* Princeton, NJ: Princeton University Press, 1996.

Heilbron, J. L. *Electricity in the 17th and 18th Centuries: A Study of Early Modern Physics.* Berkeley: University of California Press, 1979.

Herschel, J. F. W. *Preliminary Discourse on the Study of Natural Philosophy.* London: Longman, Rees, Orme, Brown & Green; John Taylor, 1830.

Hobbes, T. *Leviathan, or The Matter, Forme, and Power of a Common-Wealth Ecclesiasticall and Civill.* London: Andrew Crooke, 1651.

Holton, G. "Subelectrons, Presuppositions, and the Millikan-Ehrenhaft Dispute." *Historical Studies in the Physical Sciences* 9 (1978): 161–224.

Hossenfelder, S. *Lost in Math: How Beauty Leads Physics Astray.* New York: Basic Books, 2018.

Jahren, H. *Lab Girl.* New York: Knopf, 2016.

Jasanoff, S. "Transparency in Public Science: Purposes, Reasons, Limits." *Law and Contemporary Problems* 69 (2006): 21–45.

Johnson, G. *Strange Beauty: Murray Gell-Mann and the Revolution in Twentieth-Century Physics.* New York: Knopf, 1999.

Keller, A. "Has Science Created Technology?" *Minerva* 22 (1984): 160–82.

Kennefick, D. "Testing Relativity from the 1919 Eclipse—A Question of Bias." *Physics Today* 62, no. 3 (2009): 37–42.

Keynes, J. M. "Newton, the Man." In *The Royal Society Newton Tercentenary Celebrations 15–19 July 1946,* edited by the Royal Society (Great Britain), 27–34. Cambridge: Cambridge University Press, 1947.

Kitcher, P. *Science in a Democratic Society.* Amherst, NY: Prometheus Books, 2011.

Kragh, H. *Quantum Generations: A History of Physics in the Twentieth Century.* Princeton, NJ: Princeton University Press, 1999.

Kuhn, T. S. "The Function of Dogma in Scientific Research." In *Scientific Change: Historical Studies in the Intellectual, Social, and Technical Conditions for Scientific Discovery and Technical Invention, from Antiquity to the Present,* edited by A. C. Crombie, 347–69. New York: Basic Books, 1963.

Kuhn, T. S. *The Road since Structure: Philosophical Essays, 1970–1993, with an Autobiographical Interview.* Chicago: University of Chicago Press, 2000.

Kuhn, T. S. *The Structure of Scientific Revolutions.* 4th ed. Chicago: University of Chicago Press, 2012.

Kuhn, T. S. "What Are Scientific Revolutions?" In *The Probabilistic Revolution.* Vol. 1. Edited by L. Krüger, L. J. Daston, and M. Heidelberger, 7–22. Cambridge, MA: MIT Press, 1987.

Kulp, S. A., and B. H. Strauss. "New Elevation Data Triple Estimates of Global Vulnerability to Sea-Level Rise and Coastal Flooding." *Nature Communications* 10, 4844 (2019). doi:10.1038/s41467-019-12808-z.

Lakatos, I. "Falsification and the Methodology of Scientific Research Programmes." In *Criticism and the Growth of Knowledge*, edited by I. Lakatos and A. Musgrave, 91–196. Cambridge: Cambridge University Press, 1970.

Lamichhaney, S., F. Han, M. T. Webster, L. Andersson, B. R. Grant, and P. R. Grant. "Rapid Hybrid Speciation in Darwin's Finches." *Science* 359 (2018): 224–8.

Latour, B., and S. Woolgar. *Laboratory Life: The Construction of Scientific Facts.* 2nd ed. Princeton, NJ: Princeton University Press, 1986.

Lawrence, D. H. *Studies in Classic American Literature.* New York: Thomas Seltzer, 1923.

Leibniz, G. W. *Philosophical Essays.* Translated and edited by R. Ariew and D. Garber. Indianapolis: Hackett, 1989.

Leroi, A. M. *The Lagoon: How Aristotle Invented Science.* New York: Viking, 2014.

Lindberg, D. C. *The Beginnings of Western Science: The European Scientific Tradition in Philosophical, Religious, and Institutional Context, Prehistory to A.D. 1450.* 2nd ed. Chicago: University of Chicago Press, 2007.

Lindley, D. *Degrees Kelvin: A Tale of Genius, Invention, and Tragedy.* Washington, DC: Joseph Henry, 2004.

Macleay, W. S. *Horae Entomologicae: or, Essays on the Annulose Animals.* London: S. Bagster, 1819.

Macleay, W. S. *Illustrations of the Annulosa of South Africa; Being a Portion of the Objects of Natural History Chiefly Collected during an Expedition into the Interior of South Africa, under the Direction of Dr. Andrew Smith in the Years 1834, 1835, and 1836.* London: Smith, Elder, and Co., 1838.

Magee, B. *Popper.* London: Fontana, 1973.

Mantell, G. "The Geological Age of Reptiles." *Edinburgh New Philosophical Journal* 11 (1831): 181–5.

Manuel, F. E. *The Religion of Isaac Newton.* Oxford: Oxford University Press, 1974.

McAllister, J. W. *Beauty and Revolution in Science.* Ithaca, NY: Cornell University Press, 1996.

McGuire, J. E., and P. M. Rattansi. "Newton and the 'Pipes of Pan.'" *Notes and Records of the Royal Society of London* 21 (1966): 108–43.

Medawar, P. *Induction and Intuition in Scientific Thought.* Philadelphia: American Philosophical Society, 1969.

Mermin, D. "What Is Wrong with This Pillow?" *Physics Today* 42, no. 4 (1989): 9–11.

Morris, E. *The Ashtray (Or the Man Who Denied Reality).* Chicago: University of Chicago Press, 2018.

Mulkay, M., and G. N. Gilbert. "Putting Philosophy to Work: Karl Popper's Influence on Scientific Practice." *Philosophy of the Social Sciences* 11 (1981): 389–407.

Newton, I. *The Principia: Mathematical Principles of Natural Philosophy.* Translated by I. B. Cohen and A. Whitman. With "A Guide to Newton's *Principia*" by Cohen. Berkeley: University of California Press, 1999.

Newton, I. "Seven Statements on Religion." Keynes Ms. 6, 2002. http://www.newton project.ox.ac.uk/view/texts/normalized/THEM00006.

Oreskes, N. *The Rejection of Continental Drift: Theory and Method in American Earth Science.* Oxford: Oxford University Press, 1999.

Ospovat, D. *The Development of Darwin's Theory: Natural History, Natural Theology, and Natural Selection, 1838–1859.* Cambridge: Cambridge University Press, 1981.

Pico della Mirandola, G., *Oration on the Dignity of Man.* Translated by A. R. Caponigri. Chicago: Regnery, 1956.

Pilcher, H. "No Great Flakes." *New Scientist* 2948 (2013): 70–71.

Popper, K. *Conjectures and Refutations: The Growth of Scientific Knowledge.* London: Routledge, 1963.

Popper, K. *The Logic of Scientific Discovery.* London: Hutchinson, 1959.

Popper, K. "Natural Selection and the Emergence of Mind." *Dialectica* 32 (1978): 339–55.

Popper, K. *The Open Society and Its Enemies.* London: Routledge, 1945.

Popper, K. *Unended Quest: An Intellectual Autobiography.* Chicago: Open Court, 1976.

Proust, M. *In Search of Lost Time. Vol. II: Within a Budding Grove.* Translated by C. K. S. Moncrieff, T. Kilmartin, and D. J. Enright. New York: Modern Library, 1992.

Richardson, M. K., J. Hanken, M. L. Gooneratne, C. Pieau, A. Raynaud, L. Selwood, and G. M. Wright. "There Is No Highly Conserved Embryonic Stage in the Vertebrates: Implications for Current Theories of Evolution and Development." *Anatomy and Embryology* 196 (1997): 91–106.

Romanes, G. *A Candid Examination of Theism.* London: Trübner & Co., 1878.

Russell, B. *A History of Western Philosophy.* London: George Allen and Unwin, 1945.

Sandys, J. E. *A Short History of Classical Scholarship.* Cambridge: Cambridge University Press, 1915.

Schneider, S. H. "Confidence, Consensus and the Uncertainty Cops: Tackling Risk Management in Climate Change." In *Seeing Further: The Story of Science, Discovery and the Genius of the Royal Society,* edited by B. Bryson, 424–43. New York: HarperCollins, 2010.

Schultz, H., and D. J. Yang. *Pour Your Heart into It: How Starbucks Built a Company One Cup at a Time*. New York: Hatchette, 1997.

Scoresby, W. *An Account of the Arctic Regions with a History and Description of the Northern Whale-Fishery*. Edinburgh: Archibald Constable, 1820.

Shapin, S. *The Scientific Revolution*. Chicago: University of Chicago Press, 1996.

Smith, C., and M. N. Wise. *Energy and Empire: A Biographical Study of Lord Kelvin*. Cambridge: Cambridge University Press, 1989.

Smith, G. "Isaac Newton." In *The Stanford Encyclopedia of Philosophy*, edited by E. N. Zalta. Fall 2008 edition. http://plato.stanford.edu/archives/fall2008/entries/newton/.

Smolin, L. *The Trouble with Physics: The Rise of String Theory, the Fall of a Science, and What Comes Next*. Boston: Houghton Mifflin, 2006.

Snyder, L. J. *The Philosophical Breakfast Club: Four Remarkable Friends Who Transformed Science and Changed the World*. New York: Broadway Books, 2011.

Sprat, T. *The History of the Royal Society of London, for the Improving of Natural Knowledge*. London: T. R., 1667.

Stanley, M. *Einstein's War: How Relativity Triumphed Amid the Vicious Nationalism of World War I*. New York: Dutton, 2019.

Stanley, M. "An Expedition to Heal the Wounds of War: The 1919 Eclipse and Eddington as Quaker Adventurer." *Isis* 94 (2003): 57–89.

Stephen, L. "William Whewell." In *Dictionary of National Biography*. Vol. 60. Edited by S. Lee, 454–63. London: Macmillan, 1899.

Strickland, H. E. "On the True Method of Discovering the Natural System in Zoology and Botany." *Annals and Magazine of Natural History* 6 (1840): 184–94.

Taylor, H. O. *The Classical Heritage of the Middle Ages*. New York: Columbia University Press, 1901.

Thompson, D. W. *On Growth and Form*. Abridged ed. Edited by J. T. Bonner. Cambridge: Cambridge University Press, 1961.

Tolstoy, L. *War and Peace*. Translated by R. Pevear and L. Volokhonsky. New York: Random House, 2007.

Trout, J. D. *Wondrous Truths: The Improbable Triumph of Modern Science*. Oxford: Oxford University Press, 2016.

Turner, S. *Liberal Democracy 3.0: Civil Society in an Age of Experts*. London: SAGE Publications, 2003.

Vigors, N. A. "Observations on the Natural Affinities That Connect the Orders and Families of Birds." *Transactions of the Linnaean Society* 14 (1824): 395–517.

Wade, N. *The Nobel Duel: Two Scientists' 21-Year Race to Win the World's Most Coveted Research Prize*. New York: Anchor Doubleday, 1981.

Watson, R. A. *Cogito, Ergo Sum: The Life of René Descartes*. Jaffrey, NH: David R. Godine, 2002.

Wedgwood, C. V. *The Thirty Years War*. London: Jonathan Cape, 1938.

Weinberg, S. *Dreams of a Final Theory*. New York: Pantheon, 1992.

Weiner, J. *The Beak of the Finch*. New York: Knopf, 1994.

Westfall, R. S. *Never at Rest: A Biography of Isaac Newton*. Cambridge: Cambridge University Press, 1983.

Westfall, R. S. "Newton and the Fudge Factor." *Science* 179 (1973): 751–8.

Whewell, W. *Astronomy and General Physics Considered with Reference to Natural Theology*. Vol. 3 of *The Bridgewater Treatises on the Power, Wisdom and Goodness of God as Manifested in the Creation*. London: William Pickering, 1833.

Whewell, W. *History of the Inductive Sciences, from the Earliest to the Present Times*. London: John W Parker, 1837.

Whewell, W. *The Philosophy of the Inductive Sciences, Founded on Their History*. London: John W Parker, 1840.

Wigner, E. P. "The Unreasonable Effectiveness of Mathematics in the Natural Sciences." *Communications on Pure and Applied Mathematics* 13 (1960): 1–14.

Wilczek, F. *A Beautiful Question: Finding Nature's Deep Design*. New York: Penguin, 2015.

Wilson, E. O. *Consilience: The Unity of Knowledge*. New York: Knopf, 1998.

Wilson, E. O. *The Meaning of Human Existence*. New York: Liveright, 2014.

Wilson, P. H. *The Thirty Years War: Europe's Tragedy*. Cambridge, MA: Harvard University Press, 2009.

Wootton, D. *The Invention of Science: A New History of the Scientific Revolution*. New York: Harper, 2015.

Ziman, J. *Real Science: What It Is, and What It Means*. Cambridge: Cambridge University Press, 2000.

Zimring, J. *What Science Is and How It Really Works*. Cambridge: Cambridge University Press, 2019.

LIST OF ILLUSTRATIONS

INDEX

Italic page numbers indicate photographs and illustrations.

ABOUT THE AUTHOR

Michael Strevens is Professor of Philosophy at New York University, where since 2004 he has taught and thought about the nature of science, complex systems, the psychology of philosophy, the role of physical intuition in scientific discovery, and the nature of explanation and understanding, among other things. He was born and raised in New Zealand, graduated with a PhD in philosophy from Rutgers University, and has previously taught at Iowa State University and Stanford University. In 2017, he received a Guggenheim Fellowship. He lives in New York City with two bicycles, some skis, and a lot of books and records.